INTRODUCTION TO THE THEORY OF ABSTRACT ALGEBRAS

Richard S. Pierce

Dover Publications, Inc.
Mineola, New York

Bibliographical Note

This Dover edition, first published in 2014, is an unabridged republication of the work originally published in 1968 by Holt, Rinehart and Winston, Inc., New York, as part of the "Athena Series: Selected Topics in Mathematics." This edition is published by special arrangement with Cengage Learning, Inc., Belmont, California.

Library of Congress Cataloging-in-Publication Data

Pierce, Richard S. (Richard Scott), 1927- author.
Introduction to the theory of abstract algebras / Richard S. Pierce. — Dover edition.
pages cm — (Dover books on mathematics)
Originally published: New York : Holt, Rinehart and Winston, Inc., 1968.
Summary: "Suitable for introductory graduate-level courses and independent study, this text presents the basic definitions of the theory of abstract algebra. Following introductory material, each of four chapters focuses on a major theme of universal algebra: subdirect decompositions, direct decompositions, free algebras, and varieties of algebra. Problems and a bibliography supplement the text. "— Provided by publisher.
Includes bibliographical references and indexes.
ISBN-13: 978-0-486-78998-9 (paperback)
ISBN-10: 0-486-78998-5 (paperback)
1. Algebra, Abstract. I. Title.
QA162.P54 2014
512'.02—dc23 2014022421

Manufactured in the United States by Courier Corporation
78998501 2014
www.doverpublications.com

Preface

The theory of abstract or universal algebras provides a framework for a large part of modern algebra, and indeed of all mathematics. Familiarity with this theory should be standard equipment for all mathematicians.

Until two years ago, there was no systematic presentation of the basic results of universal algebra. When this monograph was started, it was intended to fill this gap partially. Meanwhile, the work of Cohn [9] has appeared, and Grätzer's book [21] will soon be published. Nevertheless, it seems worthwhile to go ahead with the publication of a more elementary monograph. The three books on abstract algebras will have a nonempty intersection, but each one has its own special features.

The purpose of this work is to provide an introduction to various aspects of the theory of abstract algebras. It is written to be used either as a textbook in an elementary graduate course or to be read independently by interested students. The material in the text can be covered in approximately thirty-five lectures. Thus, it is suitable either for a one quarter course (with the omission of some topics) or in a course that lasts for a semester. By supplementing the material in the text with the topics outlined in the problems, the book could be used for a full year.

The Introduction is a short exposition of the ideas of modern set theory. Initially, this chapter can be skimmed for definitions and viewpoints. Later, when the less familiar results of set theory are used, the reader can return to the proofs that are given.

Chapter 1 presents the basic definitions of the theory of abstract algebras. These are given for partial algebras or relational systems with (possibly) infinitary operations or relations. It is one feature of the book that an effort is made to present all results at their maximum level of generality. After the basic and elementary theorems, a number of examples are given in this chapter. These examples help to indicate the scope of the theory of abstract algebras.

Each of Chapters 2 through 5 focuses on a major theorem of universal algebra: Chapter 2 is devoted to the subdirect decomposition theorem and its applications. The third chapter is concerned with direct decompositions, and particularly with the uniqueness theorem of Ore. Chapter 4 deals with free extensions and free products. The basic theorem of this chapter is the existence theorem for free extensions. Finally, Chapter 5 considers varieties (or equational classes) of algebras. The fundamental theorem in this chapter is Birkhoff's characterization of varieties.

The purpose of the problems is to give examples and extensions of the text material. The problems are generally developed in small increments, with enough hints so that a student should have no difficulty in working them out for himself. Some of the material presented in these problems is of recent vintage, and a few of the results have not been previously published.

In general, there is considerable emphasis on applications of the theory. Some of the fundamental representation theorems of Boolean algebras, distributive lattices, lattice-ordered groups, and rings are presented. Lattice theory receives special attention for two reasons. First, it is the author's belief that some of the best work in lattice theory is related to the theory of abstract algebras. Second, the study of the structure of lattices is one of the most elegant of all the applications of universal algebra. Because of this bias, the monograph might appropriately be subtitled "An introduction to the theory and use of lattices."

Chapters 2 to 5 depend heavily on the material of Chapter 1. Chapter 3 requires a knowledge of section 2 in Chapter 2. Otherwise, there is little dependence between the last four chapters.

Each chapter in the book is composed of several sections, which are numbered consecutively within the chapter. The definitions, lemmas, and theorems are numbered serially, using decimal notation. For example, in section 2 of Chapter 2, Definition 2.8 is followed by Proposition 2.9, after which comes Corollary 2.10. Such a two-decimal listing is used in referring to formal statements within a given chapter. When reference is made to results or definitions in other chapters, a three-decimal system is used. For example, 4.1.5 designates Lemma 1.5 in Chapter 4.

This monograph was developed from lectures given over a number of years. The author's ideas about abstract algebras have been influenced by his teachers and his students. Most important has been the early guidance and encouragement given by R. P. Dilworth and Garrett Birkhoff. In addition, Birkhoff deserves major credit for originating many of the fundamental ideas and basic theorems of universal algebra.

Thanks are due to the University of Washington and to the National Science Foundation, the two agencies that supported the author during the last decade. Finally, it is always a pleasant duty to acknowledge the secretarial help that is so necessary for the preparation of a good manuscript. The credit for the rapid and expert execution of this manuscript belongs to Miss Betsy Dent. For her help, the author is sincerely grateful.

Richard S. Pierce

Las Cruces, New Mexico
March, 1968

Contents

INTRODUCTION TO THE THEORY OF ABSTRACT ALGEBRAS

Introduction: Set Theory

1. Introduction and Notation

The theory of abstract algebras is close enough to the foundations of mathematics that some care must be exercised in using set theory. Otherwise, there is danger of encountering contradictions similar to the familiar antinomies* of set theory. For the study of abstract algebras, the proper viewpoint toward sets lies somewhere between the formalism of axiomatic set theory and the relaxed informality that is normal for a working mathematician.

One of the principal versions of axiomatic set theory, due to Bernays, Von Neumann, and Gödel, successfully avoids the classical contradictions by making a distinction between sets and proper classes. In the axiomatic theory, "class" and "set" are primitive notions. For the informal use of set theory, it is desirable to interpret these concepts. Unfortunately, the intuitive distinction between them is not clear cut.

Roughly speaking, classes are collections of objects that satisfy some prescribed conditions. Sets are also classes. However, they are classes that consist of specific objects, as opposed to totalities of things that satisfy some property. For example, the class of all natural numbers is a set, whereas the class of all groups is not a set. Sometimes sets are described as "small" classes, but this is deceptive since sets can be very large indeed. It would perhaps be more accurate to say that the proper classes (that is, the classes that are not sets) are so enormous that the intuition is incapable of assigning any magnitude to them.

In addition to the concepts of class and set, the membership relation $\in$ is also primitive in axiomatic set theory. The axioms are formulated in terms of these undefined notions using the first-order predicate calculus (that is, the language of sentences built up by using *and*, *or*, *not*, and the quantifiers *for all* and *there exists*). These axioms include such assertions as *extensionality*: ($a \in A$ if and only if $a \in B$) implies $A = B$. The existence of unordered

* The Russell paradox is perhaps the best-known example: Consider the set S of all sets which are not members of themselves; is S a member of itself?

pairs is postulated. That is, for any two elements or sets a and b, there is a set $\{a, b\}$ such that $x \in \{a, b\}$ if and only if $x = a$ or $x = b$. The ordered pair $\langle a, b\rangle$ can then be defined as $\{a, \{a, b\}\}$. There are enough axioms concerning the existence of classes in order to guarantee that if $P(x)$ is a first-order formula with one free variable x, then a class A exists such that $a \in A$ if and only if $P(a)$. In most axiomatic formulations of set theory, there is no characterization of those classes which are sets. However, there is an axiom that proclaims the existence of an infinite set (the "axiom of infinity") and several axioms which assert that sets result from the application of certain constructions to sets. For instance, the class of all subsets of a set is a set, the image of a set under a mapping is a set, and the Cartesian product of sets is a set. Moreover, any subclass of a set is a set. In practice, the usual way of showing that a class C is a set is to prove that C can be obtained by performing some sequence of these admissible operations on given sets.

It is not our intention to give an exposition of axiomatic set theory. Too much formality could obscure the essentially simple ideas of the theory of abstract algebraic systems. Nevertheless, we will try not to use set theory in a sloppy way. It should be possible to justify all set theoretical arguments in terms of any of the axiomatic formalizations of set theory.

Sets will usually be denoted by capital Latin letters. Exceptions to this rule are ordinal and cardinal numbers, relations, functions, and (sometimes) operations. As usual, bracket symbols $\{\ \ \}$ are used to indicate set formation. The symbol $\varnothing$ will be used to represent the empty set.

Let A, B, and I be sets, and $\{C_i \mid i \in I\}$ a family of sets indexed by I.

(1.1) $A \subseteq B$ or $B \supseteq A$ means that A is a subset of B. The fact that A is a proper subset of B will be indicated by writing $A \subset B$ or $B \supset A$.

(1.2) $A \cup B$ and $A \cap B$ respectively denote the union and intersection of A and B. Denote by $A - B$ the set of elements in A that do not belong to B.

(1.3) $\bigcup_{i\in I} C_i$ and $\bigcap_{i\in I} C_i$ designate the union and intersection of the sets C_i. If $\mathfrak{C}$ is a set of sets, we will occasionally write $\bigcup\mathfrak{C}$ and $\bigcap\mathfrak{C}$ for the union and intersection of all sets in $\mathfrak{C}$.

(1.4) $A \times B$ is the product of A and B, that is, the set of all ordered pairs $\langle a, b\rangle$ with $a \in A$, $b \in B$. The ordered pair $\langle a, b\rangle$ is formally defined to be the set $\{a, \{a, b\}\}$.

(1.5) $\mathbf{X}_{i\in I} C_i$ represents the Cartesian product of the sets C_i. If $I = \{0, 1, \cdots, k-1\}$, then this product is denoted by $C_0 \times C_1 \times \cdots \times C_{k-1}$. It is customary to define $\mathbf{X}_{i\in I} C_i$ to be the set of all functions c from I to $\bigcup_{i\in I} C_i$ such that $c(i) \in C_i$ for all $i \in I$. With this definition, the notation $C_0 \times C_1$ is ambiguous, since it can mean either the set of all ordered pairs $\langle c_0, c_1\rangle$ with $c_0 \in C_0$ and $c_1 \in C_1$, or else a set of functions defined on $\{0, 1\}$. The former definition must take precedence since functions are defined by means of sets or ordered pairs. Fortunately, in practice this ambiguity causes no confusion.

(1.6) A^B denotes the set of all functions defined on B with values in A. Note that $A^{\varnothing} = \{\varnothing\}$. The notation A^B is occasionally used when A is a proper class. The meaning is the same, but of course A^B may not be a set in this case.

(1.7) $\mathbf{P}(A)$ stands for the power set of A, that is, the set of all subsets of A.

2. Mappings

A (binary) *relation* on a set A to a set B is a subset of $A \times B$. If $R \subseteq A \times B$ is a relation, the *domain* of R is

$$\mathscr{D}(R) = \{a \in A \mid \langle a, b\rangle \in R \text{ for some } b \in B\}.$$

A relation $\varphi \subseteq A \times B$ is called a *partial mapping* of A to B if $\langle a, b\rangle \in \varphi$ and $\langle a, c\rangle \in \varphi$ implies $b = c$. Moreover, φ is a *mapping* or a *function* if φ is a partial mapping such that $\mathscr{D}(\varphi) = A$.

This definition lacks something of the intuitive idea that the word "mapping" conveys. Essentially, we are identifying a mapping with its graph. However, the definition is exact and convenient for all mathematical uses.

If φ is a partial mapping of A to B and $a \in \mathscr{D}(\varphi)$, then the *value* of φ at a is denoted by $\varphi(a)$. Thus,

$$\varphi(a) = b \quad \text{if and only if} \quad \langle a, b\rangle \in \varphi.$$

For any subset A_1 of A, it is customary to define

$$\varphi(A_1) = \{\varphi(a) \mid a \in A_1 \cap \mathscr{D}(\varphi)\}.$$

In this way, every partial mapping of A to B induces a mapping of $\mathbf{P}(A)$ to $\mathbf{P}(B)$. There is generally no confusion when the same symbol is used to denote a mapping of A to B and the induced mapping of $\mathbf{P}(A)$ to $\mathbf{P}(B)$.

Let φ be a partial mapping of A to B. In addition to the mapping of $\mathbf{P}(A)$ to $\mathbf{P}(B)$ induced by φ, there is an equally important mapping of $\mathbf{P}(B)$ to $\mathbf{P}(A)$ determined by φ. For any subset B_1 of B, define

$$\varphi^{-1}(B_1) = \{a \in \mathscr{D}(\varphi) \mid \varphi(a) \in B_1\}.$$

2.1 PROPOSITION. Let φ be a mapping of A to B. In the following statements A_1 and A_2 are subsets of A, B_1 and B_2 are subsets of B, $\{A_i \mid i \in I\}$ is a set of subsets of A, and $\{B_i \mid i \in I\}$ is a set of subsets of B.

(a) $\varphi^{-1}(\varphi(A_1)) \supseteq A_1$. (b) $\varphi(\varphi^{-1}(B_1)) \subseteq B_1$. (c) If $\varphi(A) = B$, then $\varphi(\varphi^{-1}(B_1)) = B_1$. (d) If $A_1 \subseteq A_2$, then $\varphi(A_1) \subseteq \varphi(A_2)$. (e) If $B_1 \subseteq B_2$, then $\varphi^{-1}(B_1) \subseteq \varphi^{-1}(B_2)$. (f) $\varphi(\bigcup_{i\in I}A_i) = \bigcup_{i\in I}\varphi(A_i)$. (g) $\varphi^{-1}(\bigcup_{i\in I}B_i) = \bigcup_{i\in I}\varphi^{-1}(B_i)$. (h) $\varphi^{-1}(\bigcap_{i\in I}B_i) = \bigcap_{i\in I}\varphi^{-1}(B_i)$. (i) $\varphi^{-1}(B_2 - B_1) = \varphi^{-1}(B_2) - \varphi^{-1}(B_1)$.

If R is a relation on A to B, and if $A_1 \subseteq A$, then the *restriction of R to A_1* is defined by

$$R \upharpoonright A_1 = \{\langle a, b\rangle \mid \langle a, b\rangle \in R, a \in A_1\} = R \cap (A_1 \times B).$$

Evidently, if φ is a partial mapping, $(\varphi \upharpoonright A_1)(a) = \varphi(a)$ for all $a \in A_1 \cap \mathscr{D}(\varphi)$.

If $\{R_i \mid i \in I\}$ is a set of relations on A to B, then $R = \bigcup_{i\in I}R_i$ is obviously a relation on A to B. In general, R need not be a partial mapping, even if all R_i are partial mappings. However, the following lemma records one case in which the union of partial mappings is a partial mapping.

2.2 LEMMA. Let $\{\varphi_i \mid i \in I\}$ be a set of partial mappings from A to B. Suppose that for each i and j in I, there exists $k \in I$ such that $\varphi_k \supseteq \varphi_i \cup \varphi_i$. Then $\bigcup_{i\in I}\varphi_i$ is a partial mapping from A to B.

The conditions of this lemma are satisfied in particular if for each i and j either $\varphi_i \supseteq \varphi_j$ or $\varphi_j \supseteq \varphi_i$.

The *composition of mappings* will be denoted by the symbol $\circ$, or by juxtaposition, depending on which notation is clearer. That is, if φ maps A to B and ψ maps B to C, then the composite mapping $\psi \circ \varphi$ (or $\psi\varphi$), defined by $\psi \circ \varphi(a) = \psi(\varphi(a))$, maps A to C. If F is a set of mappings from A to B, and ψ is a mapping from B to C, denote $\{\psi \circ \varphi \mid \varphi \in F\}$ by $\psi \circ F$. Similarly define $G \circ \varphi$, and more generally $G \circ F$, where F and G are sets of mappings.

Evidently, composition of mappings is associative. In other words, $\chi \circ (\psi \circ \varphi) = (\chi \circ \psi) \circ \varphi$, where φ, ψ, and χ are mappings between suitable sets.

A partial mapping φ of A to B is called *one-to-one* if $\langle a_1, b\rangle \in \varphi$ and $\langle a_2, b\rangle \in \varphi$ implies $a_1 = a_2$. In other words, $\varphi(a_1) = \varphi(a_2)$ implies $a_1 = a_2$. The partial mapping φ is *onto* if $\varphi(\mathscr{D}(\varphi)) = B$.

If φ is partial mapping of A to B, which is one-to-one, then the inverse of φ is defined by

$$\varphi^{-1} = \{\langle b, a\rangle \mid \langle a, b\rangle \in \varphi\}.$$

The assumption that φ is one-to-one is exactly the condition that will imply φ^{-1} is a partial mapping. The definition of φ^{-1} given above assigns a meaning to $\varphi^{-1}(B_1)$ (for $B_1 \subseteq B$), which is consistent with the notation given previously.

There is a useful criterion for a mapping to be one-to-one and an analogous condition for being onto. These criteria are stated in terms of compositions of mappings and the notion of the *identity transformation* of a set A. Define.

$$I_A = \{\langle a, a\rangle \mid a \in A\}.$$

Clearly, I_A is a mapping of A to A, such that $I_A(a) = a$ for all $a \in A$. The conditions for a mapping to be one-to-one or onto can now be given as follows.

2.3 PROPOSITION. Let φ be a mapping of the set A to the set B.

(a) φ is one-to-one if and only if there is a mapping ψ of B to A such that $\psi\varphi = I_A$.

(b) φ is onto if and only if there is a mapping ψ of B to A such that $\varphi\psi = I_B$.

Let us prove (a). Suppose that the mapping ψ exists satisfying $\psi\varphi = I_A$. It $\varphi(a_1) = \varphi(a_2)$, then $a_1 = I_A(a_1) = \psi\varphi(a_1) = \psi(\varphi(a_1)) = \psi(\varphi(a_2)) = \psi\varphi(a_2) = I_A(a_2) = a_2$. Therefore, φ is one-to-one. Conversely, suppose that φ is one-to-one. If $A = \varnothing$, let $\psi = \varnothing$. Suppose that $A \neq \varnothing$. Choose $a_0 \in A$ arbitrarily. By definition

$$\psi = \{\langle b, a\rangle \mid \langle a, b\rangle \in \varphi\} \cup \{\langle b, a_0\rangle \mid b \in B - \varphi(A)\}.$$

It is easily verified that $\psi\varphi = I_A$. The proof of (b) is similar, except that the *axiom of choice* is needed to establish the existence of a mapping ψ such that $\varphi\psi = I_B$. Specifically, if φ is onto, then for each $b \in B$, $A_b = \{a \in A \mid \varphi(a) = b\}$ is not empty. Thus, by the axiom of choice, there is a mapping ψ of B to $\bigcup_{b\in B} A_b \subseteq A$ such that $\psi(b) \in A_b$ for all $b \in B$. Obviously, $\varphi\psi = I_B$.

If φ is a mapping of A into B, which is both one-to-one and onto, then the partial mapping φ^{-1} introduced above is a mapping. In this case, $\psi = \varphi^{-1}$ is the unique mapping that satisfies conditions (a) and (b) of Proposition 2.3.

The axiom of choice that was used in the proof of Proposition 2.3 is one of the most important hypotheses underlying modern mathematics. There are various ways in which to state this axiom. One of the simplest is the following: Let $\{A_i \mid i \in I\}$ be any set of nonempty sets; then the Cartesian product $\mathbf{X}_{i\in I} A_i$ is nonempty. In this book we will use the axiom of choice and its consequences freely.

The axiom of choice yields two principles that are of great importance in giving mathematical proofs. One of these (the well-ordering principle) will be described in the next section. The second consequence of the axiom of choice is a powerful tool for mathematical proofs.

2.4 ZORN'S LEMMA. Let $\mathfrak{A}$ be a nonempty set of sets such that the union of every simply ordered (by inclusion) nonempty subset of $\mathfrak{A}$ is again a member of $\mathfrak{A}$. Then $\mathfrak{A}$ contains a maximal set. That is, there is a set $A \in \mathfrak{A}$ such that A is not a proper subset of any other set of $\mathfrak{A}$.

The deduction of Zorn's lemma from the axiom of choice is somewhat involved. The argument is well known so that we need not present it here.

3. Well-Ordered Sets

Let W be a set, and suppose that $<$ is a (binary) relation on W to W. Assume that $<$ satisfies the following two conditions:

(i) $<$ is antisymmetric, that is, $x < y$ and $y < x$ cannot both occur;

(ii) Every nonempty subset S of W contains a smallest element, that is, there exists an element $x \in S$ such that $x < y$ holds for all $y \in S$ with $y \neq x$; then $<$ is called a *well ordering* of W. The pair $\langle W; < \rangle$ is called a *well-ordered set*. Sometimes symbols other than $<$ are used to designate the well ordering of a set W, but this is unusual. Consequently, it generally causes no confusion to speak of a well-ordered set W, rather than $\langle W; < \rangle$.

It is not hard to see that (i) and (ii) imply all of the usual properties of a well ordering. From (i) it follows that $x < x$ is impossible. Also, if $x < y$ and $y < z$, then applying (i) and (ii) to the set $\{x, y, z\}$ yields the required transitivity property $x < z$. Finally, if x and y are distinct elements of W, then using (ii) with $S = \{x, y\}$ gives $x < y$ or $y < x$, so that $<$ is a total ordering.

As usual, the expression $y > x$ will be used as a synonym for $x < y$, and $x \leqq y$ (or $y \geqq x$) means that either $x < y$ or $x = y$. By (i), $x \leqq y$ and $y \leqq x$ imply $x = y$.

A subset S of a well-ordered set W is called a *segment* of W if $v \leqq w$ and $w \in S$ implies $v \in S$. Plainly, W is a segment of itself. If $S \subset W$, and w is the smallest element of $W - S$, then it is easy to see that $S = \{u \in W \mid u < w\}$. An obvious but useful observation about segments is that any union of segments of W is a segment of W.

For the statement of the next result, it is convenient to introduce another definition. Let φ be a partial mapping from a well ordered set W to the well-ordered set V. We will say that φ *preserves order*, or is an *order-preserving* mapping if $u < v$ in $\mathscr{D}(\varphi)$ implies $\varphi(u) < \varphi(v)$. Note that if φ preserves order, then φ is one-to-one, and φ^{-1} is also order preserving. Moreover, the composition of order-preserving mappings is a mapping that preserves order.

3.1 LEMMA. Let V and W be well-ordered sets. Suppose that φ and ψ are partial mappings of V to W such that

(i) φ and ψ preserve order.

(ii) $\mathscr{D}(\varphi) = S$ and $\mathscr{D}(\psi) = T$ are segments of V, with $S \subseteq T$.

(iii) $\varphi(S)$ and $\psi(T)$ are segments of W.

Then $\varphi \subseteq \psi$.

Proof. If $\varphi \nsubseteq \psi$, then $R = \{u \in S \mid \varphi(u) \neq \psi(u)\}$ is not empty. Thus, $v \in R$ exists satisfying $v \leqq u$ for all $u \in R$. Suppose that $\varphi(v) > \psi(v)$. If

$u < v$, then $u \notin R$, so that $\varphi(u) = \psi(u) < \psi(v)$. If $u \geqq v$, then $\varphi(u) \geqq \varphi(v) > \psi(v)$. Consequently, $\psi(v) \notin \varphi(S)$. However, this is impossible because $\varphi(S)$ is an interval that contains $\{w \mid w \leqq \varphi(v)\}$. In the same way the inequality $\varphi(v) < \psi(v)$ leads to a contradiction. Thus, the assumption $\varphi \nleqq \psi$ cannot hold.

If V and W are well-ordered sets, write $V \prec W$ if there is an order-preserving mapping of V onto a segment of W. Define $V \sim W$ if there is an order-preserving mapping of V onto W.

3.2 PROPOSITION. Suppose that U, V, and W are well-ordered sets.

(a) $U \prec U$;
(b) $U \prec V$ and $V \prec W$ implies $U \prec W$; $U \sim V$ and $V \sim W$ implies $U \sim W$;
(c) $U \prec V$ and $V \prec U$ if and only if $U \sim V$;
(d) if V is a segment of W, and $V \sim W$, then $V = W$.

Proof. The statements (a) and (b) follow easily from the definitions. The statement (d) is a consequence of 3.1. In fact, if φ is an order-preserving mapping of W onto its segment V, then $\varphi = I_W$ by 3.1. Hence, $V = W$. This observation is also used in the proof of (c) which follows. If $U \prec V$ and $V \prec U$, there exist order-preserving mappings φ of U onto a segment of V and ψ of V onto a segment of U. Then $\varphi \circ \psi$ is an order-preserving mapping of V onto a segment of itself, so that $\varphi \circ \psi = I_V$. Therefore, $\varphi(U) = V$, and $U \sim V$. The converse part of (c) is an immediate consequence of the fact that if φ is an order-preserving mapping of U onto V, then φ^{-1} is an order-preserving mapping of V onto U.

3.3 PROPOSITION. Let V and W be well-ordered sets. Then either $V \prec W$, or $W \prec V$.

Proof. Let $\mathscr{G}$ be the set of all order-preserving partial mappings ψ of V to W such that

(i) $S = \mathscr{D}(\psi)$ is a segment of V, and
(ii) $\psi(S)$ is a segment of W.

Let $\varphi = \bigcup\{\psi \mid \psi \in \mathscr{G}\}$. By 3.1 and 2.2, φ is a partial mapping from V to W. Moreover, $T = \mathscr{D}(\varphi) = \bigcup\{\mathscr{D}(\psi) \mid \psi \in \mathscr{G}\}$ is a segment of V, and $\varphi(T) = \bigcup\{\psi(\mathscr{D}(\psi)) \mid \psi \in \mathscr{G}\}$ is a segment of W. Clearly, $a < b$ implies $\varphi(a) < \varphi(b)$. Therefore, $\varphi \in \mathscr{G}$. If $T = V$, then $V \prec W$. If $\varphi(T) = W$, then $W \prec V$. Suppose that $T \subset V$, and $\varphi(T) \subset W$. Let v be the smallest element of $V - T$, and choose w to be the smallest element of $W - \varphi(T)$. Define $\varphi_1 = \varphi \cup \{\langle v, w\rangle\}$. It is clear that φ_1 is an order-preserving mapping of a segment of V onto a segment of W. Therefore, $\varphi_1 \subseteq \varphi$. However, this contradicts the definition of φ_1.

It is evident that any subset of a well-ordered set W is well ordered by the restriction of the ordering of W.

3.4 LEMMA. Let V be a subset of the well-ordered set W. Then $V \prec W$.

Proof. By Proposition 3.3, either $V \prec W$, or $W \prec V$. Suppose that $W \prec V$. Let φ be an order-preserving mapping from W to a segment of V. We prove that $\varphi(a) \geqq a$ for all $a \in W$, from which it follows that $W \sim V$ (and, therefore, $V \prec W$). In fact, if $v \in V$, then $v \leqq \varphi(v) \in \varphi(W)$ implies that $v \in \varphi(W)$, since $\varphi(W)$ is a segment of V. Thus, $\varphi(W) = V$. If $\varphi(c) < c$ for some c, then there is an element $b \in W$ such that $\varphi(b) < b$ and $\varphi(a) \geqq a$ for all $a < b$. Since φ is order preserving and $\varphi(W)$ is a segment of V, this implies that $\varphi(\{a \in W \mid a < b\}) \supseteq \{c \in V \mid c < b\}$. Consequently, $\varphi(b) \geqq b$, a contradiction. Thus, $\varphi(a) \geqq a$ for all $a \in W$.

If V and W are well-ordered sets, then there is a well ordering of the set product $V \times W$ defined by

$$\langle v_0, w_0 \rangle < \langle v_1, w_1 \rangle$$

if $v_0 < v_1$ or $v_0 = v_1$ and $w_0 < w_1$. The set $V \times W$ with this ordering relation is called the *lexicographic product* of V and W.

We conclude by recording a basic theorem, which is one of the major consequences of the axiom of choice.

3.5 WELL-ORDERING THEOREM. If S is any set, then there is a relation $<$ on S such that $\langle S; < \rangle$ is a well-ordered set.

4. Ordinal Numbers

Ordinal numbers are defined to be sets; specifically, each ordinal number is the set of all smaller ordinal numbers, beginning with the empty set $\varnothing$ as the smallest ordinal. The finite ordinals are denoted as usual by $0, 1, 2, \cdots$. Thus,

$$0 = \varnothing, 1 = \{0\}, 2 = \{0, 1\}, \cdots.$$

The first transfinite ordinal is designated by ω. In other words, ω is the set of all finite ordinals:

$$\omega = \{0, 1, 2, \cdots\}.$$

We will denote infinite ordinal numbers by small Greek letters.

It is possible to give a more formal definition of ordinal numbers.

4.1 DEFINITION. An ordinal number is a set ξ such that

(i) Every element of ξ is a set.
(ii) If $\eta \in \xi$, then $\eta \subset \xi$.
(iii) The relation $\in$ well orders ξ.

The class of all ordinal numbers is denoted by "Ord."

It can easily be verified that the description of the finite ordinals given above satisfies the conditions of 4.1. The statement that each ordinal number is the set of all smaller ordinals is meaningful only when the relation "smaller" is defined for ordinal numbers. It will be clear from 4.4 below that the three most "natural" ways of ordering the ordinal numbers are equivalent, and if the term "smaller" refers to this ordering, then it is correct to say that each ordinal is the set of all smaller ordinal numbers.

4.2 LEMMA. Let ξ be an ordinal number. Then (a) every $\eta \in \xi$ is an ordinal number, and (b) every segment of ξ in the ordering defined by $\in$ is an ordinal number.

Proof. (a) If $\eta \in \xi$, then $\eta \subset \xi$ by 4.1(ii). Therefore, every element of η is a set and $\in$ well orders η. Moreover, $\zeta \in \eta$ implies $\zeta \in \xi$. Hence $\zeta \subset \xi$ by 4.1(ii) again. Consequently, $\nu \in \zeta$ and $\zeta \in \eta$ imply $\nu \in \eta$ by 4.1(iii). Therefore, $\zeta \subseteq \eta$. (b) Obviously, any segment S of ξ satisfies 4.1(i) and 4.1(iii). If $\zeta \in S$, then $\zeta \in \xi$, so that $\zeta \subset \xi$ by 4.1(ii). Thus, $\nu \in \zeta$ implies $\nu \in \xi$. Since S is a segment of ξ and $\zeta \in S$, it follows that $\nu \in \zeta$. This shows that if $\zeta \in S$, then $\zeta \subset S$, so that S satisfies 4.1(ii).

4.3 LEMMA. Let ξ and η be ordinal numbers. Then exactly one of the following relations holds

$$\xi \in \eta, \qquad \xi = \eta, \qquad \eta \in \xi.$$

Proof. Suppose that $\xi \cap \eta \subset \xi$. Choose $\xi_1 \in \xi - \xi \cap \eta$ so that $\zeta \in \xi - \xi \cap \eta$ implies $\xi_1 = \zeta$ or $\xi_1 \in \zeta$. Such a ξ_1 exists since $\in$ well orders ξ. If $\zeta \in \xi_1$, then $\zeta \in \xi$ because $\xi_1 \in \xi$ implies $\xi_1 \subset \xi$. Therefore, $\xi_1 \neq \zeta$ and $\xi_1 \notin \zeta$, so that $\zeta \in \xi \cap \eta$. Thus, $\xi_1 \subseteq \xi \cap \eta$. Suppose that $\zeta \in \xi \cap \eta$. Then $\zeta \neq \xi_1$ and $\xi_1 \notin \zeta$ (since otherwise $\xi_1 \in \xi \cap \eta$). Consequently, $\zeta \in \xi_1$. Therefore $\xi \cap \eta \subseteq \xi_1$. This shows that $\xi_1 = \xi \cap \eta$. In other words, $\xi \cap \eta \in \xi - \xi \cap \eta$ whenever $\xi \cap \eta \subset \xi$. Similarly, if $\xi \cap \eta \subset \eta$, then $\xi \cap \eta \in \eta - \xi \cap \eta$. Since $(\xi - \xi \cap \eta) \cap (\eta - \xi \cap \eta) = \varnothing$, it follows that either $\xi \cap \eta = \xi$, or $\xi \cap \eta = \eta$. If $\xi \cap \eta = \xi$ and $\xi \cap \eta \subset \eta$, then $\xi \in \eta$. If $\xi \cap \eta = \eta$ and $\xi \cap \eta \subset \xi$, then $\xi \in \eta$. The only remaining possibility is that $\xi = \eta$. The conclusion that the relations $\xi \in \eta$, $\xi = \eta$, and $\eta \in \xi$ are mutually exclusive follows from 4.1(iii) and the definition of well ordering.

4.4 COROLLARY. If ξ and η are ordinal numbers then the following conditions are equivalent:

(i) $\eta \in \xi$.
(ii) η is a proper segment of ξ in the ordering defined by $\in$.
(iii) $\eta \subset \xi$.

Proof. If $\eta \in \xi$, and $\zeta \in \eta$, then $\zeta \in \xi$. Thus, η is a segment of ξ. Since $\eta \in \xi - \eta$, the segment η is proper. Therefore, (i) implies (ii). Obviously,

(ii) implies (iii). If $\eta \subset \xi$, then $\xi \notin \eta$ because $\xi \notin \xi$. Thus, $\eta \in \xi$ by 4.3. Hence, (iii) implies (i).

This corollary provides strong evidence that inclusion is the most natural ordering for ordinal numbers. As is customary, we will write $\xi < \eta$ instead of $\xi \subset \eta$, and $\xi \leqq \eta$ in place of $\xi \subseteq \eta$.

The discussion preceding 4.1 shows how to construct finite ordinal numbers. We wish to prove that Ord contains much more than the finite ordinals.

4.5 LEMMA. The union of any set of ordinal numbers is an ordinal number.

Proof. It is only necessary to show that if $\{\xi_i \mid i \in I\} \subseteq \text{Ord}$ (where I is a set), then $\xi = \bigcup_{i \in I} \xi_i$ satisfies 4.1(iii). Let $\varnothing \neq S \subseteq \xi$. Then $S \cap \xi_i \neq \varnothing$ for some $i \in I$. Since ξ_i satisfies 4.1(iii), there is an $\eta \in S \cap \xi_i$ such that $\eta \leqq \zeta$ for all $\zeta \in S \cap \xi_i$. Suppose that $\zeta \in S$. If $\zeta \in \xi_i$, then $\eta \leqq \zeta$ by the choice of η. If $\zeta \notin \xi_i$, then $\xi_i \leqq \zeta$ by 4.3. Therefore, since $\eta \leqq \xi_i$, it follows that $\eta \leqq \zeta$.

4.6 LEMMA. If $\xi \in \text{Ord}$, then $\xi \cup \{\xi\} \in \text{Ord}$. For any ordinal η, either $\eta \leqq \xi$, or $\xi \cup \{\xi\} \leqq \eta$.

Proof. It is clear that $\xi \cup \{\xi\}$ satisfies conditions (i) and (ii) of 4.1. If $\varnothing \neq S \subseteq \xi \cup \{\xi\}$, then either $S \cap \xi \neq \varnothing$, or $S = \{\xi\}$. In the first case, there is an $\eta \in S \cap \xi$ such that $\eta \leqq \zeta$ for all $\zeta \in S \cap \xi$. Since $\eta \leqq \xi$, it follows that η is the smallest element of S. If $S = \{\xi\}$, then ξ is obviously the smallest element of S. To prove the last statement, note that either $\eta \leqq \xi$, or else $\xi < \eta$. In the latter case, $\xi \in \eta$, so that $\xi \cup \{\xi\} \subseteq \eta$. Hence, $\xi \cup \{\xi\} \leqq \eta$.

The ordinal number $\xi \cup \{\xi\}$ is usually denoted by $\xi + 1$. In case ξ is finite, this notation is consistent with the familiar "addition of one." For example, $1 = 0 + 1 = \varnothing \cup \{\varnothing\} = \{\varnothing\}$, $2 = 1 + 1 = \{\varnothing\} \cup \{\{\varnothing\}\} = \{\varnothing, \{\varnothing\}\}$ and so forth. The last statement of 4.6 amounts to saying that there is no ordinal between ξ and $\xi + 1$.

4.7 PROPOSITION. Let W be a well-ordered set. Then there is a unique ordinal number ξ such that $W \sim \xi$.

Proof. Let $\mathfrak{S} = \{\xi \in \text{Ord} \mid \zeta \prec W\}$. Define $\eta = \bigcup \mathfrak{S}$. By 4.5, $\eta \in \text{Ord}$. Therefore, $\eta + 1 \in \text{Ord}$ by 4.6. Hence, $\eta + 1 \succ W$ or $W \succ \eta + 1$ by 3.3. If $\eta + 1 \prec W$, then $\eta + 1 \subseteq \eta$, which implies that $\eta \in \eta$. However, this is impossible by 4.3. Therefore, $W \prec \eta + 1$. That is, $W \sim S$, where S is a segment of $\eta + 1$. By 4.2(b), $W \sim \xi$ for some ordinal number ξ. If $W \sim \xi_0$ and $W \sim \xi_1$, then $\xi_0 \sim \xi_1$. By 4.3 and 4.4, either ξ_0 is a segment of ξ_1, or ξ_1 is a segment of ξ_0. Either possibility implies that $\xi_0 = \xi_1$ by 3.2(d).

If W is a well-ordered set, the *order type* $\tilde{W}$ of W is defined to be the unique ordinal number ξ that satisfies $W \sim \xi$. By 3.4, $V \subseteq W$ implies $\tilde{V} \leqq \tilde{W}$.

5. Cardinal Numbers

If A and B are sets, then A and B are called *equipollent* if there is a one-to-one mapping of A onto B.

A *cardinal number* is an ordinal number α with the property that if $\xi < \alpha$, then ξ is not equipollent to α.

From this definition and 4.3, it follows that no two distinct cardinal numbers are equipollent.

If A is any set, then there is a unique cardinal number α such that α is equipollent to A. Indeed, by 3.5, A can be well ordered. If this is done, then $A \sim \xi$ for some ordinal number ξ (by 4.7). Plainly, $A \sim \xi$ implies that A is equipollent to ξ. Therefore, the set S of all ordinal numbers that are equipollent to A is not empty. It is obvious that the smallest element of S is a cardinal number α that is equipollent to A. If A were equipollent to cardinal numbers α and β, then α and β would be equipollent. Therefore $\alpha = \beta$ as we noted above. We will denote the unique cardinal number which is equipollent to A by $|A|$.

The cardinal number $|A|$ is called the *cardinality* (or *cardinal number*) of the set A. It is evident that if α is a cardinal number, then $|\alpha| = \alpha$. In general, $|\xi| \leqq \xi$ for any ordinal number ξ. If ξ is a finite ordinal number, then $|\xi| = \xi$. That is, every finite ordinal number is a cardinal number. So is the first infinite ordinal number ω. However, if ξ is an infinite ordinal number, then $|\xi + 1| = |\xi|$. In fact, in the axiomatic development of set theory, this property is taken as the definition of an infinite ordinal number.

The class of all cardinal numbers, being a subclass of the well-ordered class of all ordinal numbers, is well ordered (by inclusion). Using this observation and Proposition 5.3(a) below, the method employed in the proof of 3.3 shows that there is a unique one-to-one, order-preserving function on the class of all ordinal numbers to the class of all infinite cardinal numbers. This is the familiar "aleph" function. It is customary to denote the image of an ordinal ξ under this function by $\aleph_\xi$. In particular, $\aleph_0 = \omega$.

There are useful arithmetics defined for both ordinal and cardinal numbers. These are obtained by introducing operations of addition, multiplication, and exponentiation. In the case of the finite ordinals and cardinals, these arithmetical operations are the same as the usual ones. However, the cardinal operations on infinite cardinal numbers are quite different from the ordinal operations. Except for the notation $\xi + 1$ denoting the successor of the ordinal ξ, no use will be made of ordinal operations in what follows. However, operations with cardinal numbers are used in later chapters. Accordingly, the definitions and basic properties of these operations will be reviewed here.

Let α and β be cardinal numbers. Let A be any set which is equipollent to α, and let B be a set which is equipollent to β, and such that $A \cap B = \varnothing$. Define†

$$\alpha + \beta = |A \cup B|,$$

$$\alpha \times \beta = |A \times B|,$$

$$\alpha^\beta = |A^B|.$$

It is easy to see that these definitions do not depend on the choice of A and B. For example, if C is equipollent to α and D is equipollent to β, then C^D is equipollent to A^B, so that $|C^D| = |A^B|$.

We will use cardinal arithmetic to obtain estimates of the cardinality of various sets. The following elementary results are used frequently for estimating cardinalities.

5.1 Proposition. Let α and β be cardinal numbers, and suppose that A and B are sets such that $|A| \leqq \alpha$ and $|B| \leqq \beta$. Assume that I is an index set satisfying $|I| \leqq \alpha$, and that $\{B_i \mid i \in I\}$ is a set of sets such that $|B_i| \leqq \beta$ for all $i \in I$.

(a) $|A \cup B| \leqq \alpha + \beta$

(b) $|A \times B| \leqq \alpha \times \beta$

(c) $|A^B| \leqq \alpha^\beta$

(d) $|\bigcup_{i \in I} B_i| \leqq \alpha \times \beta$.

It follows easily from the definitions that cardinal addition and multiplication satisfy the commutative, associative, and distributive laws:

$$\alpha + \beta = \beta + \alpha,$$

$$\alpha \times \beta = \beta \times \alpha,$$

$$(\alpha + \beta) + \gamma = \alpha + (\beta + \gamma),$$

$$\alpha \times (\beta \times \gamma) = (\alpha \times \beta) \times \gamma,$$

$$\alpha \times (\beta + \gamma) = \alpha \times \beta + \alpha \times \gamma.$$

Moreover, the rules of exponents are valid:

$$\alpha^{\beta+\gamma} = \alpha^\beta \times \alpha^\gamma$$

$$(\alpha \times \beta)^\gamma = \alpha^\gamma \times \beta^\gamma$$

$$(\alpha^\beta)^\gamma = \alpha^{\beta\times\gamma}.$$

† The notation α^β can cause confusion, since the same expression is used to designate the set of all mappings from β to α. However, this ambiguity will cause no confusion in the following chapters.

Finally, it is easy to show that these operations are monotone in the following senses: if α, β, and γ are cardinal numbers such that $\alpha \leqq \beta$, then

$$\alpha + \gamma \leqq \beta + \gamma,$$
$$\alpha \times \gamma \leqq \beta \times \gamma$$
$$\alpha^\gamma \leqq \beta^\gamma,$$
$$\gamma^\alpha \leqq \gamma^\beta.$$

As a subject in itself, the arithmetic of infinite cardinal numbers is not particularly interesting. The following result shows why this is so.

5.2 PROPOSITION. If α and β are infinite cardinal numbers, then

(a) $\alpha + \beta = \max\{\alpha, \beta\}$.
(b) $\alpha \times \beta = \max\{\alpha, \beta\}$.

We will not prove this fundamental theorem of set theory, except to note that the results follow from their special cases $\alpha + \alpha = \alpha$, $\alpha \times \alpha = \alpha$. Indeed, if $\beta \leqq \alpha$ for example, then $\alpha \leqq \alpha + \beta \leqq \alpha + \alpha = \alpha$, and $\alpha \leqq \alpha \times \beta \leqq \alpha \times \alpha = \alpha$.

Certain rules of exponents play an important part in the study of abstract algebras.

5.3 PROPOSITION. Let α, β, and γ be cardinal numbers.

(a) If $\alpha \geqq 2$, then $\alpha^\beta > \beta$.
(b) If $\beta \geqq 1$, then $\alpha^\beta \geqq \alpha$.
(c) If α is infinite, and $\alpha = \gamma^\beta$, then $\alpha^\beta = \alpha$.
(d) If $\beta \geqq 1$ and there is a mapping φ of β to α such that $\bigcup_{\eta<\beta}\varphi(\eta) = \alpha$, then $\alpha^\beta > \alpha$.

Proof. (a) If $\alpha^\beta \leqq \beta$, then it is possible to index the set of all mappings from β to α by β, say $\{\chi_\xi \mid \xi < \beta\}$ (where repetitions are allowed). Since $\alpha \geqq 2$, $\alpha - \{\chi_\xi(\xi)\} \neq \varnothing$ for every $\xi < \beta$. Consequently, there is a function ψ defined on β such that $\psi(\eta) \in \alpha - \{\chi_\xi(\xi)\}$ for every $\xi < \beta$. Plainly, ψ cannot be χ_ξ, since $\psi(\xi) \neq \chi_\xi(\xi)$. Thus, $\alpha^\beta \leqq \beta$ leads to a contradiction. Therefore $\alpha^\beta > \beta$.

The statement (b) of 5.3 is a special case of the last monotonicity rule mentioned above.

(c) If $\alpha = \gamma^\beta$ where β is infinite, then $\alpha^\beta = (\gamma^\beta)^\beta = \gamma^{\beta\times\beta} = \gamma^\beta = \alpha$ by 5.1. If β is finite, then γ must be infinite in order for α to be infinite. In this case, (c) follows from 5.2 by induction on β.

(d) Suppose that $\alpha^\beta \leqq \alpha$. Then the set of all mappings from β to α can be indexed by α. Let $\{\chi_\xi \mid \xi < \alpha\}$ be such an indexing. Since α is a cardinal number and $\varphi(\eta) < \alpha$ for all $\eta < \beta$, it follows that $|\varphi(\eta)| < \alpha$. Therefore, if $\eta < \beta$, then $|\{\chi_\xi(\eta) \mid \xi < \varphi(\eta)\}| < \alpha$. In particular, the set

$U_\eta = \alpha - \{\chi_\xi(\eta) \mid \xi < \varphi(\eta)\}$ is not empty. For $\eta < \beta$, define $\psi(\eta)$ to be the least element of U_η. Then ψ is a mapping from β to α, so that $\psi = \chi_\xi$ for some $\xi < \alpha = \bigcup_{\eta<\beta}\varphi(\eta)$. Consequently, there exists $\eta < \beta$ such that $\xi < \varphi(\eta)$. However, this gives a contradiction, because $\psi(\eta) \in U_\eta$ and $\chi_\xi(\eta) \notin U_\eta$. Therefore, the supposition that $\alpha^\beta \leqq \alpha$ must be false. Consequently, $\alpha^\beta > \alpha$.

An infinite cardinal number α is called *singular* if there is a set $V \subseteq \alpha$ such that $|V| < \alpha$, and $\bigcup V = \alpha$. Otherwise, α is called *regular*. Evidently ω is regular. Moreover, if $\alpha = \aleph_{\xi+1}$, then $V \subseteq \alpha$, $|V| < \alpha$ implies that $|\beta| \leqq \aleph_\xi$ for all $\beta \in V$, and $|V| \leqq \aleph_\xi$. Therefore, $|\bigcup V| \leqq \aleph_\xi \cdot \aleph_\xi = \aleph_\xi$. In particular, $\bigcup V \neq \alpha$. Hence α is regular. An example of a singular cardinal number is furnished by $\aleph_\omega$. In fact, it is easy to see that $\aleph_\omega = \bigcup_{n<\omega}\aleph_n$.

Notes to the Introduction

The purpose of this introductory chapter is to present an informal exposition of the von Neumann-Berneys-Gödel version of set theory. More complete developments of this topic can be found in the first part of Gödel's monograph [20] or the appendix of Kelley's book [32]. In particular, Kelley shows how Zorn's lemma and the well-ordering principle can be deduced from the axiom of choice. The Zermelo-Frankel theory could equally well be used as a basis for the development of abstract algebra.

[*1*]

Basic Concepts

1. Algebras and Relational Systems

The purpose of this section is to introduce the principal object of interest in this book, the notion of an abstract algebra.

1.1 DEFINITION. Let A be a set, and suppose that τ is an ordinal number. A *τ-ary partial operation on A* is a partial mapping of A^τ to A. That is, F is a τ-ary partial operation if F maps $D \subseteq A^\tau$ to A. If $\mathscr{D}(F) = A^\tau$, then F is called a *τ-ary operation on A*.

1.2 EXAMPLES. (a) $\tau = 0$. A *zero-ary partial operation* is either the empty set or it is a mapping from $A^0 = \{\varnothing\} = \{0\}$ into A. In the latter case, it is an operation. Thus, the effect of a zero-ary operation on A is to specify an element of A, and it is common practice to denote a zero-ary operation by listing the element that it selects.

(b) $\tau = 1$. A *unary partial operation* on A is essentially a mapping from a subset of A to A, since A^1 can be identified with A by the correspondence $a \to a(0)$.

(c) $\tau = 2$. A *binary partial operation* on A corresponds to a mapping from a subset of $A \times A$ to A when the natural identification $a \leftrightarrow \langle a(0), a(1)\rangle$ is made between A^2 and $A \times A$. Examples of such binary operations are common. For instance, the multiplication and addition operations in a ring are binary. Frequently, composition notation is used to designate binary operations. For example, the operation of addition is denoted by $a + b$ rather than $+(a, b)$.

1.3 DEFINITION. Let A be a set, and suppose that τ is an ordinal number. A *τ-ary relation on A* is a subset of A .

1.4 EXAMPLES. (a) $\tau = 0$. The only 0-ary relations on A are $\varnothing = 0$ and $A^0 = \{0\} = 1$.

(b) $\tau = 1$. The *unary relations* on A are identified with subsets of A by the natural correspondence between A^1 and A.

(c) $\tau = 2$. The *binary relations* on A correspond to sets of ordered pairs by the identification of A^2 with $A \times A$. Generally, we will consider a binary relation as a set of ordered pairs.

If R is a binary relation on A, it is customary to write aRb to indicate that $\langle a, b\rangle \in R$.

The correspondence between A^2 and $A \times A$ mentioned in Example 1.2(c) can be generalized to an identification of $A^{\tau+1}$ with $A^\tau \times A$. The correspondence is given by

$$\mathbf{a} \leftrightarrow \langle \mathbf{a} \upharpoonright \tau, \mathbf{a}(\tau)\rangle,$$

where $\mathbf{a}$ is in $A^{\tau+1}$. Under this identification, a τ-ary partial operation F on A is associated with a $(\tau + 1)$-ary relation on A. Indeed, $F \subseteq A^\tau \times A$ goes into a subset of $A^{\tau+1}$, and $\mathbf{a} \in F$ is equivalent to $\mathbf{a} \upharpoonright \tau \in \mathscr{D}(F)$ and $F(\mathbf{a} \upharpoonright \tau) = \mathbf{a}(\tau)$. We will sometimes find it useful to think of partial operations as relations. Whenever such phrases as "since partial operations are special cases of relations" occur, the implied reference is to the identification of a τ-ary partial operation with a $(\tau + 1)$-ary relation given by the correspondence between $A^\tau \times A$ and $A^{\tau+1}$ which has just been described.

1.5 Definition. Let J be any set, and suppose that $\tau \in (\mathrm{Ord})^J$. A *partial algebra of similarity type* τ is a system* $\mathbf{A} = \langle A; F_j\rangle_{j\in J}$ consisting of a set A, together with a set of partial operations on A, indexed by J, such that F_j is $\tau(j)$-ary. If all of the partial operations F_j for $j \in J$ are operations (that is, if $\mathscr{D}(F_j) = A^{\tau(j)}$ for all $j \in J$), then A is called an *abstract algebra of similarity type* τ.

1.6 Definition. Let J be any set, and suppose that $\tau \in (\mathrm{Ord})^J$. A *relational system of similarity type* τ is a system $\mathbf{D} = \langle D; R_j\rangle_{j\in J}$ consisting of a set D, together with a set of relations on D, indexed by J, such that R_j is $\tau(j)$-ary.

It is customary to say that two algebras or relational systems are *similar* if they have the same similarity type. Moreover, the terminology introduced in 1.5 and 1.6 is frequently abbreviated. For example, it is usual to refer to "an algebra of type τ" rather than an abstract algebra of similarity type τ. Another common practice is to use the same symbol to denote a relational system and the set of its elements. For example, it is standard procedure to speak of a group G, meaning the group whose set of elements is G. This practice will be avoided in this book. Instead, we adopt once and for all the convention of denoting relational systems, partial algebras, and abstract algebras by boldface capital letters, with the corresponding standard capital letter used to designate the set of elements in the system. For example, if A

* Throughout this book, the term "system" will be used in referring to an ordered pair whose first element is a set A and whose second element is a mapping F with domain J, which takes its values in the classes of operations, partial operations, or relations on A. In using this terminology and the elaborate notation $\langle A; F_j\rangle_{j\in J}$, we are simply following tradition.

appears in some discussion of an algebra **A**, it will be assumed without mention that A is the set of elements of **A**. In other words, $\mathbf{A} = \langle A; F_j \rangle_{j \in J}$.

In most discussions of abstract algebras or relational systems, attention is fixed on one algebra, or on a class of algebras of a given (but possibly arbitrary) similarity type. It is usually not important what set J is used for the definition of this type. Indeed, J serves only to index the relations or partial operations of the algebras. Of course, in a discussion of several relational systems of the same type, it is important that the same indexing be used for all systems under consideration. However, the nature of the index set is immaterial. It will usually be assumed that the set J is an ordinal number ρ. The relations or operations are then indexed by the ordinals $\xi < \rho$. Occasionally, the operations or relations of a system are indexed simply by listing them in a definite order. For example, this is done for common algebras such as groups, rings, and lattices. It can be assumed that listing the operations is equivalent to indexing by an ordinal number. There are cases in which it is not natural to index the operations by an ordinal number. For instance, if a vector space over a field **F** is considered as a group with operators, then the unary operations corresponding to the various scalar multiplications are most conveniently indexed by the set of elements of **F**.

Since any τ-ary partial operation is a special case of a $(\tau + 1)$-ary relation, it follows that a partial algebra of type $\tau \in (\mathrm{Ord})^{\rho}$ is also a relational system of type σ, where $\sigma(\xi) = \tau(\xi) + 1$ for all $\xi < \rho$. One may wonder then why we do not restrict our attention to relational systems. It turns out that the theory of algebras and partial algebras is much richer than that of relational systems. Moreover, the generality that is gained by considering relational systems rather than partial algebras is more apparent than real. (See Problem 3 at the end of this chapter.) Consequently, most of our attention will be directed toward partial algebras and algebras. Occasionally, however, it is more natural to state a definition or theorem for relational systems than for partial algebras. In such cases, we will of course deal with relational systems.

Most familiar algebraic systems can be considered as abstract algebras, partial algebras, or relational systems. The following examples indicate the wide scope of this concept.

1.7 EXAMPLE. Probably the simplest kind of system that interests algebraists is a semigroup. A *semigroup* is an abstract algebra of type $\tau = \{\langle 0, 2 \rangle\}$ (thus, $\mathcal{D}(\tau) = 1$ and $\tau(0) = 2$). The binary operation of a semigroup is usually designated by composition notation: $F_0(\langle x, y \rangle) = x \times y$. The property that distinguishes semigroups from other abstract algebras with a single binary operation is associativity, that is, the condition that $x \times (y \times z) = (x \times y) \times z$ for all elements x, y, and z.

1.8 EXAMPLE. A *group* is a particular kind of abstract algebra of type $\{\langle 0, 2 \rangle, \langle 1, 0 \rangle, \langle 2, 1 \rangle\}$. The binary operation in a group is associative and, as in the case of a semigroup, it is denoted by composition. This operation is called *group multiplication*. The zero-ary operation picks out an element e, satisfying the usual conditions for an identity element, namely, $e \times x = x \times e = x$ for every element x. The unary operation is the group inverse, generally denoted $F_2(x) = x^{-1}$. The characteristic property of the inverse is expressed by the equations $x \times x^{-1} = x^{-1} \times x = e$ for all x. By definition then, an abstract algebra $\mathbf{G} = \langle G; \times, e, ^{-1}\rangle$ of type $\{\langle 0, 2\rangle, \langle 1, 0\rangle, \langle 2, 1\rangle\}$ is a group if and only if the binary operation $(\times)$ is associative, $x \times e = e \times x = x$ for all $x \in G$, and $x \times x^{-1} = x^{-1} \times x = e$ for all $x \in G$.

It is well known that groups can be defined using only the binary multiplication operation. From this viewpoint, a group is a special kind of semigroup. Many familiar algebraic systems can be considered as abstract algebras in more than one way. When applying the general theory of abstract algebras to particular systems, it is important to keep in mind what type is being attributed to the algebra, since some of the concepts of the general theory have meanings that depend on the type. For example, if a group is considered as an abstract algebra of type $\{\langle 0, 2\rangle\}$, then the term "subalgebra" (which will be defined in section 4) means sub-semigroup. On the other hand, if the same group is viewed as an algebra of type $\{\langle 0, 2\rangle, \langle 1, 0\rangle, \langle 2, 1\rangle\}$, then the concept of a subalgebra is the same as the usual notion of a subgroup.

1.9 EXAMPLE. A *ring* $\mathbf{R} = \langle R; +, 0, -, \times\rangle$ is an abstract algebra of type $\{\langle 0, 2\rangle, \langle 1, 0\rangle, \langle 2, 1\rangle, \langle 3, 2\rangle\}$. With respect to its first three operations $+, 0, -$, $\mathbf{R}$ is an Abelian group, that is, a group that satisfies the commutative law $x + y = y + x$ for all x and y. The binary operation of multiplication $(\times)$ is associative and distributive with respect to addition $(+)$: $x \times (y + z) = x \times y + x \times z$ and $(x + y) \times z = x \times z + y \times z$ for all x, y, z in R. A ring can also be considered as an abstract algebra of type $\{\langle 0, 2\rangle, \langle 1, 2\rangle\}$ with only the binary operations of addition and multiplication.

1.10 EXAMPLE. *Modules* over a ring $\mathbf{R}$ can be considered as abstract algebras. The operations of a module $\mathbf{A}$ consist of one binary operation $+$, and a set of unary operations F_α corresponding to the elements $\alpha \in R$. Thus, $\mathbf{A} = \langle A; +, F_\alpha\rangle_{\alpha \in R}$. As a rule, composition is used to denote the unary operations of a module, that is, $F_\alpha(x) = \alpha x$. In terms of this notation, the following conditions constitute a set of axioms for modules:

(i) $\mathbf{A}$ forms a commutative group with respect to the operation $+$.
(ii) $\alpha(x + y) = \alpha x + \alpha y$ for all $x, y \in A$ and all $\alpha \in R$.
(iii) $(\alpha + \beta)x = \alpha x + \beta x$ for all $x \in A$ and all $\alpha, \beta \in R$.
(iv) $\alpha(\beta x) = (\alpha \times \beta)x$ for all $x \in A$ and all $\alpha, \beta \in R$.

It should be emphasized that the operations of $\mathbf{R}$ do not occur among the operations of $\mathbf{A}$. However, the addition and multiplication in $\mathbf{R}$ influence the structure of $\mathbf{A}$ through their appearance in the identities (iii) and (iv). Note that there is one identity of the form (ii) for each $\alpha \in R$. Similarly, every choice of α and β gives rise to identities of the form (iii) and (iv).

1.11 EXAMPLE. A *field* can be conveniently considered as a partial algebra $\mathbf{F} = \langle F; +, 0, -, \times, 1, ^{-1}\rangle$. The type is $\{\langle 0, 2\rangle, \langle 1, 0\rangle, \langle 2, 1\rangle, \langle 3, 2\rangle$ $\langle 4, 0\rangle, \langle 5, 1\rangle\}$. All of the partial operations except the inverse $(^{-1})$ are operations, and $\mathscr{D}(^{-1}) = F - \{0\}$. In fact, by definition, a partial algebra $\langle F; +, 0, -, \times, 1, ^{-1}\rangle$ is a field exactly when $\langle F; +, 0, -, \times\rangle$ is a ring, and $\langle F - \{0\}; \times, 1, ^{-1}\rangle$ is an Abelian group.

1.12 EXAMPLE. One extremely important kind of relational system that is not an abstract algebra is a *partially ordered set*. A binary relation $\leq$ on a set A is called a *partial ordering* of A if

(i) $x \leq x$ for all $x \in A$,
(ii) $x \leq y$ and $y \leq z$ implies $x \leq z$, and
(iii) $x \leq y$ and $y \leq x$ implies $x = y$.

The system $\mathbf{A} = \langle A; \leq\rangle$ is called a partially ordered set. The similarity type of a partially ordered set is $\{\langle 0, 2\rangle\}$.

1.13 EXAMPLE. Abstract algebras with infinitary operations are not uncommon. Let $\mathscr{F}$ be a family of subsets of a set X, such that $\mathscr{F}$ is closed under countable unions, and under the operation of taking complements: $A \to A^c = X - A$. Then $\mathscr{F}$ determines an abstract algebra $\langle \mathscr{F}; \bigcup_{n<\omega}, {}^c\rangle$ of type $\{\langle 0, \omega\rangle, \langle 1, 1\rangle\}$. These algebras are special cases of a class of algebraic systems known as *ω-complete Boolean algebras*, which will be defined and studied later (see Examples 2.4.2, 4.1.8, and 5.3.7).

1.14 EXAMPLE. A *small category* can be defined as a partial algebra $\mathbf{H} = \langle H; \times, e_i\rangle_{i \in I}$, where $\times$ is a binary partial operation and, for each $i \in I$, e_i is an element of H (that is, an element distinguished by a zero-ary operation) satisfying the following conditions:

(i) H is the disjoint union of the sets $H_{i,j} = \{x \in H \mid \langle e_i, x\rangle \in \mathscr{D}(\times), \langle x, e_j\rangle \in \mathscr{D}(\times)\}$, where $i, j \in I$.
(ii) $\mathscr{D}(\times) = \bigcup_{i,j,k \in I}\{\langle x, y\rangle \mid x \in H_{i,j}, y \in H_{j,k}\}$.
(iii) If $x \in H_{i,j}$ and $y \in H_{j,k}$, then $x \times y \in H_{i,k}$.
(iv) If $x \in H_{i,j}$, $y \in H_{j,k}$, and $z \in H_{k,l}$, then $(x \times y) \times z = x \times (y \times z)$.
(v) $e_i \in H_{i,i}$.
(vi) If $x \in H_{i,j}$, then $e_i \times x = x$ and $x \times e_j = x$.

Notice that since H is a partial algebra, H and I must be sets. When this restriction is dropped, the axioms given above define a *category*.

As a particular example of a small category, let H be the set of all (rectangular) matrices with entries in some field. Let the binary operation $\times$ be matrix multiplication, with $x \times y$ defined if and only if the number of columns of x is the same as the number of rows of y. Let I be the set of all natural numbers, and for $i \in I$, let e_i be the $i \times i$ identity matrix. It is easy to show that the partial algebra $\langle H; \times, e_i \rangle_{i \in I}$ defined in this way is a small category. Plainly, $H_{i,j}$ is the set of all $i \times j$ matrices with entries in the field.

1.15 EXAMPLE. Let M be a metric space. Define an ω-ary partial operation L on M as follows:

$$\mathscr{D}(L) = \{x \in M \mid \lim_{n \to \infty} x(n) \text{ exists}\},$$

$$L(x) = \lim_{n \to \infty} x(n) \qquad \text{for } x \in \mathscr{D}(L).$$

It is clear that this partial operation uniquely determines the topology on M. Thus, the concept of a partial algebra embraces part of metric space topology. Replacing sequences by nets, it is possible to extend this example to more general spaces. In later examples, we will call $\mathbf{M} = \langle M; L \rangle$ the *limit algebra* of the metric space M.

2. Homomorphisms

The concept of a homomorphism is of the greatest importance in all branches of algebra. It is not surprising that this notion also occupies a central place in the theory of abstract algebras.

2.1 DEFINITION. Let $\mathbf{A} = \langle A; R_\xi \rangle_{\xi < \rho}$ and $\mathbf{B} = \langle B; S_\xi \rangle_{\xi < \rho}$ be similar† relational systems. A mapping $\varphi: A \to B$ is called a *homomorphism of* $\mathbf{A}$ *to* $\mathbf{B}$ if $\varphi \circ R_\xi \subseteq S_\xi$ for all $\xi < \rho$ (where $\varphi \circ R_\xi$ denotes $\{\varphi \circ \mathbf{a} \mid \mathbf{a} \in R_\xi\}$). If φ is also one-to-one, then φ is called a *monomorphism*. If φ is onto, and if $\varphi \circ R_\xi = S_\xi$ for all $\xi < \rho$, then φ is called an *epimorphism*.‡ If φ is both a monomorphism and an epimorphism, then it is called an *isomorphism*. If there exists an isomorphism of $\mathbf{A}$ to $\mathbf{B}$, then $\mathbf{A}$ is said to be *isomorphic* to $\mathbf{B}$. The statement "$\mathbf{A}$ is isomorphic to $\mathbf{B}$" is expressed symbolically by $\mathbf{A} \cong \mathbf{B}$.

A homomorphism of $\mathbf{A}$ to itself is called an *endomorphism*, and an isomorphism of $\mathbf{A}$ to $\mathbf{A}$ is an *automorphism*.

Notation. The set of all homomorphisms of $\mathbf{A}$ to $\mathbf{B}$ will be denoted by $\mathrm{Hom}(\mathbf{A}, \mathbf{B})$.

Definition 2.1 applies in particular to partial algebras and algebras. For such systems, the definition of a homomorphism can be given a more familiar form.

† It should be emphasized that homomorphisms are only defined between algebras of the same similarity type.

‡ The definitions of monomorphism and epimorphism given in 2.1 are not always equivalent to the definitions of these concepts given in the theory of categories, even in the case that $\mathbf{A}$ and $\mathbf{B}$ are abstract algebras.

2.2 PROPOSITION. Let $\mathbf{A} = \langle A; F_\xi \rangle_{\xi<\rho}$, $\mathbf{B} = \langle B; G_\xi \rangle_{\xi<\rho}$ be similar partial algebras. A mapping φ from A to B is a homomorphism if and only if $\mathbf{a} \in \mathscr{D}(F_\xi)$ implies $\varphi \circ \mathbf{a} \in \mathscr{D}(G_\xi)$, and $\varphi(F_\xi(\mathbf{a})) = G_\xi(\varphi \circ \mathbf{a})$ for all $\xi < \rho$.

Proof. By the identification of operations with relations, $\mathbf{a} \in \mathscr{D}(F_\xi)$ if and only if there exists $\mathbf{f} \in F_\xi$ such that $\mathbf{a} = \mathbf{f} \restriction \tau(\xi)$. In this case $\mathbf{f}(\tau(\xi)) = F_\xi(\mathbf{a})$. If φ is in Hom(**A**, **B**), then $\mathbf{a} = \mathbf{f} \restriction \tau(\xi)$ with $\mathbf{f} \in F_\xi$ implies that $\varphi \circ \mathbf{f} \in G_\xi$, and $\varphi \circ \mathbf{a} = (\varphi \circ \mathbf{f}) \restriction \tau(\xi)$. Thus, $\varphi \circ \mathbf{a} \in \mathscr{D}(G_\xi)$. Moreover, $G_\xi(\varphi \circ \mathbf{a}) = (\varphi \circ \mathbf{f})(\tau(\xi)) = \varphi(\mathbf{f}(\tau(\xi))) = \varphi(F_\xi(\mathbf{a}))$. Conversely, suppose that for all $\xi < \rho$, $\mathbf{a} \in \mathscr{D}(F_\xi)$ implies $\varphi \circ \mathbf{a} \in \mathscr{D}(G_\xi)$, and $\varphi(F_\xi(\mathbf{a})) = G_\xi(\varphi \circ \mathbf{a})$. Let $\mathbf{f} \in F_\xi$. Then $\mathbf{a} = \mathbf{f} \restriction \tau(\xi) \in \mathscr{D}(F_\xi)$, and $\mathbf{f}(\tau(\xi)) = F_\xi(\mathbf{a})$. By hypothesis, $\varphi \circ \mathbf{a} \in \mathscr{D}(G_\xi)$ and $G_\xi(\varphi \circ \mathbf{a}) = \varphi(F_\xi(\mathbf{a}))$. Therefore, $\varphi \circ \mathbf{a} = \mathbf{g} \restriction \tau(\xi)$ for some $\mathbf{g} \in G_\xi$. Moreover, $\mathbf{g}(\tau(\xi)) = G_\xi(\mathbf{g} \restriction \tau(\xi)) = G_\xi(\varphi \circ \mathbf{a}) = \varphi(F_\xi(\mathbf{a})) = \varphi(\mathbf{f}(\tau(\xi)))$. Thus, for all $\eta \leqq \tau(\xi)$, $\varphi(\mathbf{f}(\eta)) = \mathbf{g}(\eta)$. That is, $\varphi \circ \mathbf{f} = \mathbf{g} \in G_\xi$. This proves that $\varphi \circ F_\xi \subseteq G_\xi$, so that $\varphi \in$ Hom (**A**, **B**).

2.3 EXAMPLES. (a) Let $\mathbf{A} = \langle A; \times \rangle$ and $\mathbf{B} = \langle B; \times \rangle$ be small categories (see Example 1.9). A homomorphism φ of **A** to **B** is called a *covariant functor* in the theory of categories.

(b) Let $\mathbf{M}_1 = \langle M_1, L \rangle$ and $\mathbf{M}_2 = \langle M_2, L \rangle$ be limit algebras of metric spaces (see Example 1.15). A mapping $\varphi: M_1 \to M_2$ is a homomorphism if and only if it is continuous in the topological sense, and it is an isomorphism if and only if it is a topological homeomorphism.

In the case of an algebra the definition of an epimorphism can be simplified.

2.4 PROPOSITION. Let φ be a homomorphism of the abstract algebra $\mathbf{A} = \langle A; F_\xi \rangle_{\xi<\rho}$ to a similar partial algebra $\mathbf{B} = \langle B; G_\xi \rangle_{\xi<\rho}$. Suppose that $\varphi(A) = B$. Then φ is an epimorphism and **B** is an algebra.

Proof. By hypothesis $\varphi \circ F_\xi \subseteq G_\xi$ for all ξ. To prove that φ is an epimorphism, it is necessary to show that $\varphi \circ F_\xi = G_\xi$. Let $\mathbf{g} \in G_\xi$. Define $\mathbf{b} = \mathbf{g} \restriction \tau(\xi)$. Then $\mathbf{b} \in \mathscr{D}(G_\xi)$ and $G_\xi(\mathbf{b}) = \mathbf{g}(\tau(\xi))$. Since $\varphi(A) = B$, there exists $\mathbf{a} \in A^{\tau(\xi)}$ such that $\varphi \circ \mathbf{a} = \mathbf{b}$. The hypothesis that **A** is an algebra guarantees that $\mathbf{a} \in \mathscr{D}(F_\xi)$. Thinking of F_ξ as a relation, this means that there is an $\mathbf{f} \in F_\xi$ such that $\mathbf{f} \restriction \tau(\xi) = \mathbf{a}$ and $\mathbf{f}(\tau(\xi)) = F_\xi(\mathbf{a})$. Therefore, by Proposition 2.2, $\varphi(\mathbf{f}(\tau(\xi)) = \varphi(F_\xi(\mathbf{a})) = G_\xi(\varphi \circ \mathbf{a}) = G_\xi(\mathbf{b}) = \mathbf{g}(\tau(\xi))$. This equality, together with the fact that $\varphi \circ (\mathbf{f} \restriction \tau(\xi)) = \varphi \circ \mathbf{a} = \mathbf{b} = \mathbf{g} \restriction \tau(\xi)$ implies that $\varphi \circ \mathbf{f} = \mathbf{g}$. Consequently, $\varphi \circ F_\xi = G_\xi$. Hence, φ is an epimorphism. Moreover, $\mathscr{D}(G_\xi) = \{\mathbf{g} \restriction \tau(\xi) \mid \mathbf{g} \in G_\xi\} = \{(\varphi \circ \mathbf{f}) \restriction \tau(\xi) \mid \mathbf{f} \in F_\xi\} = \varphi \circ \{\mathbf{f} \restriction \tau(\xi) \mid \mathbf{f} \in F_\xi\} = \varphi \circ \mathscr{D}(F_\xi) = \varphi \circ A^{\tau(\xi)} = B^{\tau(\xi)}$. Thus, **B** is an algebra.

2.5 LEMMA. Let **A**, **B**, and **C** be similar relational systems. Let $\varphi \in$ Hom (**A**, **B**), $\psi \in$ Hom (**B**, **C**). Then $\psi \circ \varphi \in$ Hom (**A**, **C**). Moreover, if φ and ψ are monomorphisms (epimorphisms), then $\psi \circ \varphi$ is a monomorphism (epimorphism).

This lemma is a direct consequence of Definition 2.1. In spite of its elementary nature, Lemma 2.5 has many applications. The fact that the composition of homomorphisms produces a homomorphism is the basis of the convenient diagrams, so common in mathematics. Let **A** and **B** be relational systems of the same type. The statement that φ is a homomorphism of **A** to **B** can be conveyed concisely by either of the diagrams

$$\mathbf{A} \xrightarrow{\varphi} \mathbf{B}, \quad \text{or} \quad \varphi : \mathbf{A} \longrightarrow \mathbf{B}.$$

Moreover, several homomorphisms can be represented at once schematically. For example, the meaning of

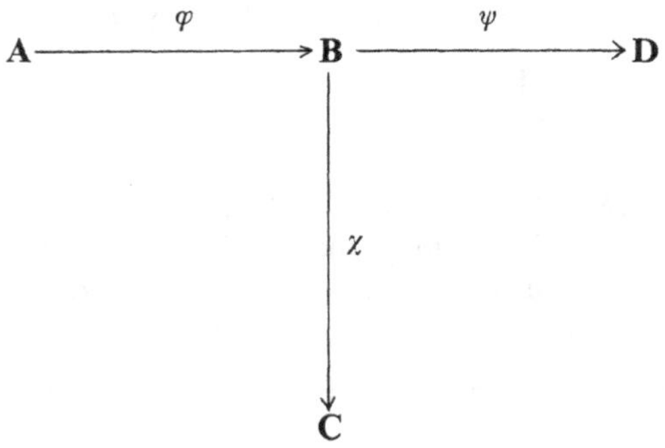

is expressed more clearly by the picture than it could be in words. The facts that φ and χ compose to give a homomorphism from **A** to **C**, and φ and ψ compose to give a homomorphism from **A** to **D** are naturally inferred from the diagram. A diagram is called *consistent* or *commutative* if, whenever there are two sequences of arrows leading from one algebra to another, the compositions of the homomorphisms along these paths produce the same homomorphism. For example

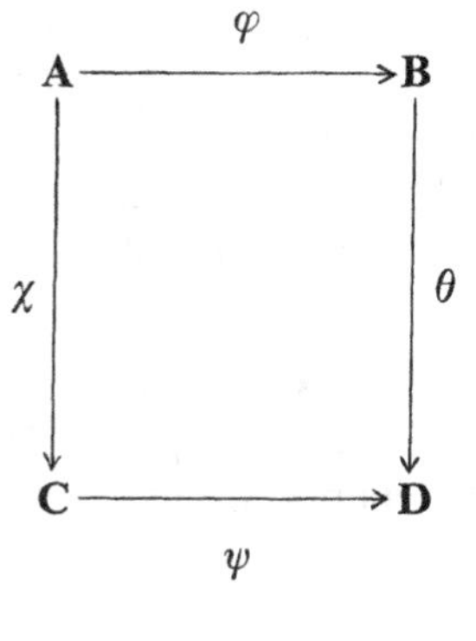

is consistent if and only if $\theta \circ \varphi = \psi \circ \chi$, while

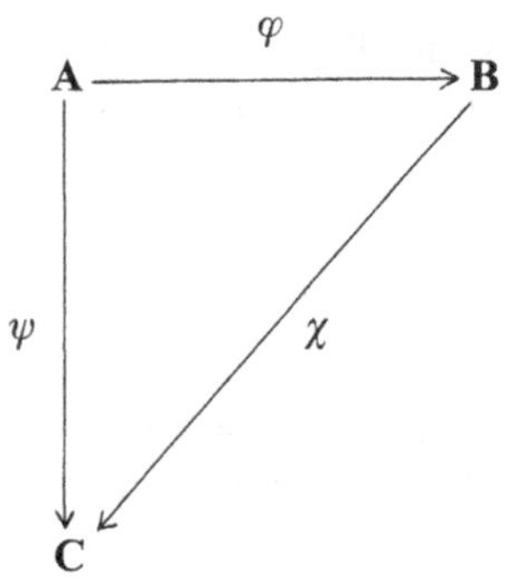

is consistent if and only if $\chi \circ \varphi = \psi$.

For any relational system $\mathbf{A} = \langle A; R_\xi \rangle_{\xi < \rho}$, the set Hom (**A**, **A**) is a semigroup with respect to the binary operation of composition. Clearly, the identity mapping I_A of A onto itself is an isomorphism of **A** to **A**, so that it belongs to Hom (**A**, **A**). We call this mapping the *identity isomorphism of* **A**.

The composition of homomorphisms is the path on which category theory enters the study of abstract algebras and relational systems. The class of all homomorphisms of relational systems of type τ forms§ a category $\mathbf{H}_\tau$ (see the definition given in Example 1.14). The binary partial operation of this category is the composition of mappings, with domain consisting of all pairs $\langle \varphi, \psi \rangle$ with $\psi \in$ Hom (**A**, **B**), $\varphi \in$ Hom (**B**, **C**), where **A**, **B**, and **C** are any relational systems of type τ. The distinguished objects of $\mathbf{H}_\tau$ are the identity isomorphisms of the various systems of type τ. The categories $\mathbf{H}_\tau$ are not small, since the class of all relational systems of type τ is not a set.

Isomorphisms can be characterized by conditions analogous to those given in Proposition 2.3 of the Introduction.

2.6 PROPOSITION. Let **A** and **B** be similar relational systems. Let $\varphi \in$ Hom (**A**, **B**). Then φ is an isomorphism if and only if there exists $\psi \in$ Hom (**B**, **A**) such that $\varphi \circ \psi = I_B$ and $\psi \circ \varphi = I_A$.

Proof. Suppose that such a ψ exists. By Proposition 2.3 found in the Introduction, φ is one-to-one and onto. Hence, φ is a monomorphism. Moreover, if $\mathbf{A} = \langle A; R_\xi \rangle_{\xi < \rho}$ and $\mathbf{B} = \langle B; S_\xi \rangle_{\xi < \rho}$, then $\varphi \circ R_\xi \subseteq S_\xi$ and $\psi \circ S_\xi \subseteq R_\xi$. Consequently, $\varphi \circ R_\xi \supseteq \varphi \circ (\psi \circ S_\xi) = (\varphi \circ \psi) \circ S_\xi = I_B \circ S_\xi = S_\xi$. Hence, $\varphi \circ R_\xi = S_\xi$. Thus, φ is also an epimorphism. It follows that

§ In order to have $\mathbf{H}_\tau$ satisfy the category axioms, we must assume that Hom (**A**, **B**) $\cap$ Hom (**A**′, **B**′) $= \emptyset$ if $\mathbf{A} \neq \mathbf{A}'$ or $\mathbf{B} \neq \mathbf{B}'$. This requires a modification of the interpretation of mappings as sets of ordered pairs.

φ is an isomorphism. Conversely, if φ is an isomorphism, then it is easy to verify that $\psi = \varphi^{-1}$ has the required properties.

2.7 COROLLARY. Let **A** and **B** be similar relational systems. If φ is an isomorphism of **A** to **B**, then φ^{-1} is an isomorphism of **B** to **A**.

3. Congruence Relations

One of the main tools in the study of the structure of a partial algebra **A** is the lattice of congruence relations of **A**. The purpose of this section is to introduce this concept, and to show how congruence relations are connected with epimorphisms.

3.1 DEFINITION. Let A be a set. A binary relation R on A is called an *equivalence relation on A* if it satisfies

(i) $a \mathrel{R} a$ for all $a \in A$.
(ii) $a \mathrel{R} b$ implies $b \mathrel{R} a$.
(iii) $a \mathrel{R} b$ and $b \mathrel{R} c$ imply $a \mathrel{R} c$.

The set of all equivalence relations on A is denoted by $E(A)$.

There are two particularly important equivalence relations on a set A:

$$I_A = \{\langle a, a\rangle \mid a \in A\}, \text{ and } U_A = A \times A.$$

If $R \in E(A)$, then $I_A \subseteq R \subseteq U_A$, so that I_A is the smallest, U_A the largest equivalence relation on A. An important property of $E(A)$ is its closure under intersections. In fact, if $\{R_i \mid i \in I\} \subseteq E(A)$, then $\bigcap_{i\in I} R_i$ obviously satisfies the conditions of 3.1; thus, $\bigcap_{i\in I} R_i \in E(A)$.

A fundamental construction in algebra is the process of passing from an equivalence relation on a set A to the set of equivalence classes of elements of A. This construction is probably familiar to the reader, but we will repeat it here, partly to establish notation.

3.2 DEFINITION. Let R be an equivalence relation on the set A. For $a \in A$, the *equivalence class of a* is

$$\bar{R}(a) = \{b \in A \mid \langle a, b\rangle \in R\}.$$

Denote by A/R the set of all equivalence classes, that is $A/R = \{\bar{R}(a) \mid a \in A\}$. The symbol $\bar{R}$ will be used to denote the mapping $a \to \bar{R}(a)$ determined by R.

The definition of an equivalence relation R on a set A leads directly to the basic properties of A/R.

3.3 PROPOSITION. Let R be an equivalence relation on the set A.

(a) $a \in \bar{R}(a)$.
(b) If $\bar{R}(a) \cap \bar{R}(b) \neq \varnothing$, then $\bar{R}(a) = \bar{R}(b)$.
(c) $\bar{R}(a) = \bar{R}(b)$ if and only if $\langle a, b\rangle \in R$.

We turn now to the definition of congruence relations.

3.4 DEFINITION. Let $\mathbf{A} = \langle A; F_\xi \rangle_{\xi<\rho}$ be a partial algebra. A *congruence relation on* $\mathbf{A}$ is an equivalence relation Γ on A which satisfies

(i) for all $\xi < \rho$, if $\mathbf{a} \in \mathscr{D}(F_\xi)$, $\mathbf{b} \in \mathscr{D}(F_\xi)$, and $\langle \mathbf{a}(\eta), \mathbf{b}(\eta) \rangle \in \Gamma$ for all $\eta < \tau(\xi)$, then $\langle F_\xi(\mathbf{a}), F_\xi(\mathbf{b}) \rangle \in \Gamma$.

The set of all congruence relations on $\mathbf{A}$ is denoted by $\Theta(\mathbf{A})$.

3.5 PROPOSITION. Let $\mathbf{A} = \langle A; F_\xi \rangle_{\xi<\rho}$ be a partial algebra. Then $\Theta(\mathbf{A})$ contains I_A and U_A. Moreover, $\Theta(\mathbf{A})$ is closed under intersection, that is, if $\{\Gamma_i \mid i \in I\} \subseteq \Theta(\mathbf{A})$, then $\bigcap_{i\in I}\Gamma_i \in \Theta(\mathbf{A})$.

Proof. It was observed above that I_A and U_A are equivalence relations on $\mathbf{A}$, and it is easily seen that I_A and U_A satisfy the condition (i) of 3.4. Hence I_A and U_A belong to $\Theta(\mathbf{A})$. Suppose that $\{\Gamma_i \mid i \in I\} \subseteq \Theta(\mathbf{A})$. Let $\mathbf{a} \in \mathscr{D}(F_\xi)$, $\mathbf{b} \in \mathscr{D}(F_\xi)$ be such that $\langle \mathbf{a}(\eta), \mathbf{b}(\eta) \rangle \in \bigcap_{i\in I}\Gamma_i$ for all $\eta < \tau(\xi)$. In particular, for each $i \in I$, and for all $\eta < \tau(\xi)$, $\langle \mathbf{a}(\eta), \mathbf{b}(\eta) \rangle \in \Gamma_i$. Therefore, since Γ_i is a congruence relation, $\langle F_\xi(\mathbf{a}), F_\xi(\mathbf{b}) \rangle \in \Gamma_i$ for all $i \in I$. That is, $\langle F_\xi(\mathbf{a}), F_\xi(\mathbf{b}) \rangle \in \bigcap_{i\in I}\Gamma_i$. Therefore, $\bigcap_{i\in I}\Gamma_i$ is a congruence relation on $\mathbf{A}$.

With each homomorphism of a partial algebra $\mathbf{A}$ there is associated a congruence relation on $\mathbf{A}$. Moreover, every congruence relation arises from a homomorphism (in fact epimorphism) of $\mathbf{A}$. It is the object of the remainder of this section to develop this correspondence between homomorphisms and congruences.

3.6 LEMMA. Let $\mathbf{A}$ and $\mathbf{B}$ be partial algebras of the same type, and let φ be a homomorphism of $\mathbf{A}$ to $\mathbf{B}$. Define

$$\Gamma_\varphi = \{\langle a, b \rangle \in A \times A \mid \varphi(a) = \varphi(b)\}.$$

Then Γ_φ is a congruence relation on $\mathbf{A}$.

Terminology. Γ_φ will be called the *kernel* of φ.

Proof. It is trivial to verify that Γ_φ is an equivalence relation on A. Let $\mathbf{A} = \langle A; F_\xi \rangle_{\xi<\rho}$, $\mathbf{B} = \langle B; G_\xi \rangle_{\xi<\rho}$. Assume $\mathbf{a} \in \mathscr{D}(F_\xi)$, $\mathbf{b} \in \mathscr{D}(F_\xi)$, and $\langle \mathbf{a}(\eta), \mathbf{b}(\eta) \rangle \in \Gamma_\varphi$ for all $\eta < \tau(\xi)$. By definition of Γ_φ, this means that $\varphi(\mathbf{a}(\eta)) = \varphi(\mathbf{b}(\eta))$ for all $\eta < \tau(\xi)$. In other words, $\varphi \circ \mathbf{a} = \varphi \circ \mathbf{b}$. Consequently, since φ is a homomorphism, $\varphi(F_\xi(\mathbf{a})) = G_\xi(\varphi \circ \mathbf{a}) = G_\xi(\varphi \circ \mathbf{b}) = \varphi(F_\xi(\mathbf{b}))$. Thus, $\langle F_\xi(\mathbf{a}), F_\xi(\mathbf{b}) \rangle \in \Gamma_\varphi$. This shows that Γ_φ is a congruence relation.

Before proving that every congruence arises from an epimorphism, we establish a very important fact relating an epimorphism to its kernel.

3.7 PROPOSITION. Let $\mathbf{A}$, $\mathbf{B}$, and $\mathbf{C}$ be similar partial algebras. Suppose that φ is an epimorphism of $\mathbf{A}$ to $\mathbf{B}$, and ψ is a homomorphism from $\mathbf{A}$ to $\mathbf{C}$. A necessary and sufficient condition for the existence of a homomorphism θ

from **B** to **C** so that the diagram

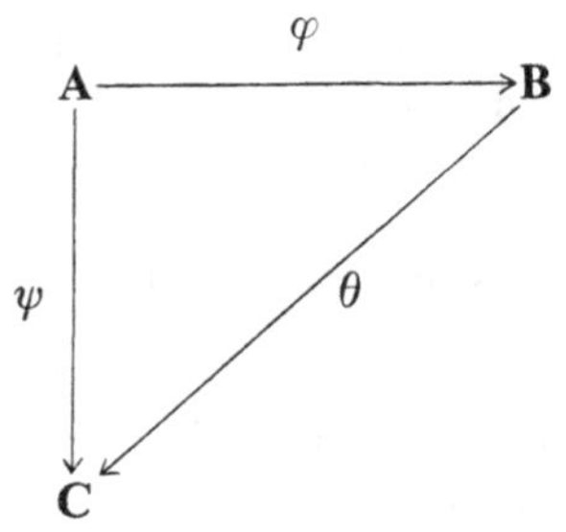

is consistent (that is, $\theta\varphi = \psi$) is that $\Gamma_\varphi \subseteq \Gamma_\psi$.

Proof. Assume that a homomorphism θ exists satisfying $\theta\varphi = \psi$. Then $\langle a_1, a_2\rangle \in \Gamma_\varphi$ implies $\varphi(a_1) = \varphi(a_2)$. Consequently, $\psi(a_1) = \theta\varphi(a_1) = \theta\varphi(a_2) = \psi(a_2)$. Hence, $\langle a_1, a_2\rangle \in \Gamma_\psi$. This shows that $\Gamma_\varphi \subseteq \Gamma_\psi$. To prove the converse, assume $\Gamma_\varphi \subseteq \varphi_\psi$. If b is an element of B, then, since φ is an epimorphism, there is an element a in A such that $\varphi(a) = b$. We wish to define $\theta(b) = \psi(a)$, but it is necessary to show that this definition does not depend on an arbitrary choice of a. The reason that it does not is that if $\varphi(a') = b = \varphi(a)$, then $\langle a, a'\rangle \in \Gamma_\varphi \subseteq \Gamma_\psi$, so that $\psi(a) = \psi(a')$. Clearly, $\theta\varphi = \psi$ by the definition of θ. It is only necessary to prove that θ is a homomorphism. Let $\mathbf{g} \in G_\xi$, where G_ξ is one of the partial operations of **B**. For convenience, consider the partial operations as relations in this proof. Since φ is an epimorphism, there exists $\mathbf{f}$ in the corresponding partial operation F_ξ of **A** such that $\varphi \circ \mathbf{f} = \mathbf{g}$. By definition, $\theta \circ \mathbf{g} = \psi \circ \mathbf{f} \in H_\xi$ (the corresponding partial operation of **C**). Thus, $\theta \circ G_\xi \subseteq H_\xi$, which is the result required to show that θ is a homomorphism.

3.8 COROLLARY. Let **A**, **B**, and **C** be similar partial algebras. Suppose that φ is an epimorphism of **A** to **B** and ψ is an epimorphism of **A** to **C**. Then there is an isomorphism θ of **B** to **C** such that $\theta\varphi = \psi$ if and only if $\Gamma_\varphi = \Gamma_\psi$.

Proof. Suppose that an isomorphism $\theta: B \to C$ exists such that $\theta\varphi = \psi$. Then $\Gamma_\varphi \subseteq \Gamma_\psi$ by 3.7. Moreover, since θ is an isomorphism, $\varphi = \theta^{-1}\psi$, so that $\Gamma_\psi \subseteq \Gamma_\varphi$ by 3.7 again. Consequently, $\Gamma_\varphi = \Gamma_\psi$. Conversely, assume that $\Gamma_\varphi = \Gamma_\psi$. By 3.7, homomorphisms $\theta_1:\mathbf{B} \to \mathbf{C}$, and $\theta_2:\mathbf{C} \to \mathbf{B}$ exist satisfying $\theta_1\varphi = \psi$, and $\theta_2\psi = \varphi$. Consequently,

$$\theta_1\theta_2\psi = \psi, \text{ and } \theta_2\theta_1\varphi = \varphi.$$

Since φ and ψ are epimorphisms, it follows from these relations that

$$\theta_1\theta_2 = I_C \text{ and } \theta_2\theta_1 = I_B.$$

By 2.6, we conclude that θ_1 is an isomorphism.

It is natural to call two epimorphisms φ and ψ *equivalent* if they have the property $\Gamma_\varphi = \Gamma_\psi$. Corollary 3.8 shows what this means in terms of the mappings of algebras of type τ.

3.9 PROPOSITION. Let $\mathbf{A} = \langle A; F_\xi \rangle_{\xi<\rho}$ be a partial algebra of type τ. Suppose that Γ is a congruence relation on $\mathbf{A}$. For $\xi < \rho$, $\bar{\Gamma} \circ F_\xi$ is a $\tau(\xi)$-ary partial operation|| on A/Γ, so that

$$\mathbf{A}/\Gamma = \langle A/\Gamma, \bar{\Gamma} \circ F_\xi \rangle_{\xi<\rho}$$

is a partial algebra of similarity type τ. Moreover, $\bar{\Gamma}$ is an epimorphism of $\mathbf{A}$ to $\mathbf{A}/\Gamma$, such that the kernel of $\bar{\Gamma}$ is Γ.

Proof. Clearly, $\bar{\Gamma} \circ F_\xi \subseteq (A/\Gamma)^{\tau(\xi)+1}$. Suppose that $\mathbf{f}$ and $\mathbf{g}$ are elements of F_ξ such that $(\bar{\Gamma} \circ \mathbf{f}) \restriction \tau(\xi) = (\bar{\Gamma} \circ \mathbf{g}) \restriction \tau(\xi)$. Then $\langle \mathbf{f}(\eta), \mathbf{g}(\eta) \rangle \in \Gamma$ for all $\eta < \tau(\xi)$. Since Γ is a congruence relation, it follows that $\langle F_\xi(\mathbf{f} \restriction \tau(\xi)), F_\xi(\mathbf{g} \restriction \tau(\xi)) \rangle \in \Gamma$, that is, $\langle \mathbf{f}(\tau(\xi)), \mathbf{g}(\tau(\xi)) \rangle \in \Gamma$. Thus,

$$\bar{\Gamma}(\mathbf{f}(\tau(\xi))) = \bar{\Gamma}(\mathbf{g}(\tau(\xi))),$$

so that, $\bar{\Gamma} \circ F_\xi$ is a partial operation. Clearly, $\bar{\Gamma}$ maps A onto A/Γ, and by definition the operations of $\mathbf{A}/\Gamma$ are $\bar{\Gamma} \circ F_\xi$. Hence, $\bar{\Gamma}$ is an epimorphism of $\mathbf{A}$ to A/Γ. Finally, by 3.3, $\bar{\Gamma}(a) = \bar{\Gamma}(b)$ if and only if $\langle a, b \rangle \in \Gamma$. Thus, the kernel of $\bar{\Gamma}$ is Γ.

The partial algebra $\mathbf{A}/\Gamma$ is called the *quotient algebra* of $\mathbf{A}$ determined by Γ. It follows from 2.4 that if $\mathbf{A}$ is an algebra, then $\mathbf{A}/\Gamma$ is also an algebra.

The results 3.8 and 3.9 show that there is a one-to-one correspondence between the equivalence classes of epimorphisms of $\mathbf{A}$ and the set $\Theta(\mathbf{A})$ of all congruence relations on $\mathbf{A}$. It might be thought desirable to extend this result to relational systems. However, it is easy to see that every equivalence relation on A determines an epimorphism of the relational system $\langle A; R_\xi \rangle_{\xi<\rho}$, and two epimorphisms correspond to the same equivalence relation if and only if they are equivalent in the sense described above. Not much information can be obtained about a relational system $\mathbf{A} = \langle A; R_\xi \rangle_{\xi<\rho}$ from the study of $E(A)$. For this reason, there is little point in trying to generalize the concept of a congruence relation to relational systems.

Combining Corollary 3.8 and Proposition 3.9, we obtain a result which in concrete instances is known as the *Homomorphism theorem.*

3.10 COROLLARY. Let $\mathbf{A}$ and $\mathbf{B}$ be similar partial algebras, and let $\varphi: \mathbf{A} \to \mathbf{B}$ be an epimorphism. Then there is an isomorphism $\theta: \mathbf{A}/\Gamma_\varphi \to \mathbf{B}$

|| Of course, we are considering $\tau(\xi)$-ary partial operations as $(\tau(\xi) + 1)$-ary relations in this proposition.

such that

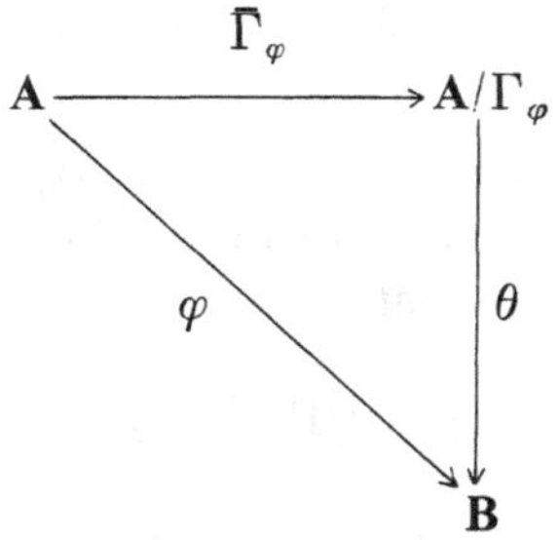

is consistent. That is, $\mathbf{B} \cong \mathbf{A}/\Gamma_\varphi$.

4. Subalgebras

The concept *subalgebra of an algebra* is very important in mathematics. This notion has a natural definition for abstract algebras, and even for partial algebras. Unfortunately, there seems to be no obvious "right way" to define subalgebras of relational systems. There is a very weak notion which has some importance.

4.1 Definition. Let $\mathbf{A} = \langle A; R_\xi \rangle_{\xi < \rho}$ and $\mathbf{B} = \langle B; S_\xi \rangle_{\xi < \rho}$ be similar relational systems. Then **B** is called a *subsystem* of **A**, and **A** is called an *extension* of **B** if $B \subseteq A$ and $S_\xi \subseteq R_\xi$ for all $\xi < \rho$.

It is obvious from this definition that **B** is a subsystem of **A** if and only if $B \subseteq A$, and the inclusion mapping I_B of B to A is a homomorphism. It is easy to show that such inclusion mappings are the prototypes of all monomorphisms, and that subsystems are related to monomorphisms much in the same way as the quotient algebras of A are related to epimorphisms. The exact statement of this relation is given in the following analogue of Corollary 3.10.

4.2 Proposition. Let **A** and **B** be similar relational systems, and let $\varphi: \mathbf{A} \to \mathbf{B}$ be a monomorphism. Then there is an extension **C** of **A** and an isomorphism $\theta: \mathbf{C} \to \mathbf{B}$ such that the diagram below is consistent.

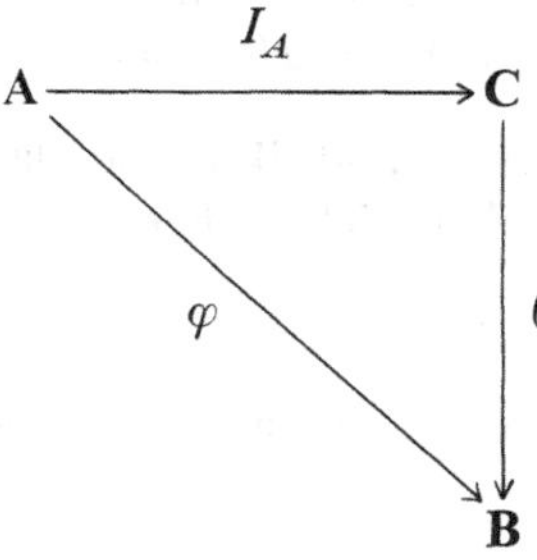

Proof. Let $\mathbf{A} = \langle A; R_\xi \rangle_{\xi < \rho}$ and $\mathbf{B} = \langle B; S_\xi \rangle_{\xi < \rho}$. Define $C = A \cup (B - \varphi(A))$, and $\theta = \varphi \cup I_{B-\varphi(A)}$. Then θ is a one-to-one mapping of C onto B. If C is equipped with relations $T_\xi = \theta^{-1} \circ S_\xi$ for $\xi < \rho$, then $\mathbf{C} = \langle C; T_\xi \rangle_{\xi < \rho}$ becomes a relational system that is similar to $\mathbf{B}$, such that θ is an isomorphism of $\mathbf{C}$ onto $\mathbf{B}$. Moreover, since $\varphi R_\xi \subseteq S_\xi$, it follows that $R_\xi = \theta^{-1} \circ \varphi \circ R_\xi \subseteq \theta^{-1} \circ S_\xi = T_\xi$ for all $\xi < \rho$. Hence, $\mathbf{C}$ is an extension of $\mathbf{A}$.

If $\mathbf{B}$ is a subsystem of the relational system $\mathbf{A}$, and if $\varphi \in \operatorname{Hom}(\mathbf{A}, \mathbf{C})$, then it is clear from Definitions 4.1 and 2.1 that $\varphi \restriction B$ is a homomorphism of $\mathbf{B}$ to $\mathbf{C}$. Moreover, if φ is a monomorphism, then $\varphi \restriction B$ is also a monomorphism. Tacit use is often made of these observations.

If $\mathbf{A} = \langle A; R_\xi \rangle_{\xi < \rho}$ is a relational system of type τ and $B \subseteq A$, then B determines a subsystem of $\mathbf{A}$ called the *restriction of* $\mathbf{A}$ *to* B, defined by

$$\mathbf{A} \restriction B = \langle B; R_\xi \cap B^{\tau(\xi)} \rangle_{\xi < \rho}.$$

Clearly, $\mathbf{A} \restriction B$ is the largest subsystem of $\mathbf{A}$ with B as its set of elements.

A trivial but useful observation is the following: if $B \supseteq C$, then $(\mathbf{A} \restriction B) \restriction C = \mathbf{A} \restriction C$. This transitivity property is an immediate consequence of the definition of restrictions.

If $\mathbf{A}$ is a partial algebra, then $\mathbf{A} \restriction B$ is also a partial algebra for any $B \subseteq A$. However, if $\mathbf{A}$ is an algebra and B is an arbitrary subset of A, then $\mathbf{A} \restriction B$ may not be an algebra. The set B must satisfy a certain condition for $\mathbf{A} \restriction B$ to be an algebra. Moreover, this condition on the subsets of A makes sense, and is important even for partial algebras.

4.3 Definition. Let $\mathbf{A} = \langle A; F_\xi \rangle_{\xi < \rho}$ be a partial algebra of similarity type τ. A subset B of A *determines a subalgebra of* $\mathbf{A}$ if the condition

$$\mathbf{b} \in \mathscr{D}(F_\xi) \cap B^{\tau(\xi)} \quad \text{implies} \quad F_\xi(\mathbf{b}) \in B$$

is satisfied. An algebra $\mathbf{B}$ of type τ is called a *subalgebra of* $\mathbf{A}$ if $B \subseteq A$, B determines a subalgebra of $\mathbf{A}$, and $\mathbf{B} = \mathbf{A} \restriction B$.

It is clear from Definition 4.3 that if $\mathbf{A}$ is an algebra, and if the subset B of A determines a subalgebra, then $\mathbf{A} \restriction B$ is an algebra.

It is customary to abuse the language of 4.3 occasionally. Thus, we will speak of the subalgebra B of $\mathbf{A}$, when we really mean a set B that determines a subalgebra of $\mathbf{A}$. Usually, however, this inexactness causes no confusion.

For any partial algebra $\mathbf{A} = \langle A; F_\xi \rangle_{\xi < \rho}$. Let $\Sigma(\mathbf{A})$ denote the set of all subsets of A that determine subalgebras of $\mathbf{A}$. It is sometimes convenient to abuse the notation by writing $\mathbf{B} \in \Sigma(\mathbf{A})$ to mean that $\mathbf{B}$ is a subalgebra of $\mathbf{A}$.

Two properties of the set $\Sigma(\mathbf{A})$ are trivial consequences of the definition. Nevertheless, they turn out to be very useful.

4.4 Proposition. Let $\mathbf{A} = \langle A; F_\xi \rangle_{\xi < \rho}$ be a partial algebra. Then $\Sigma(\mathbf{A})$ contains A. Moreover, $\Sigma(\mathbf{A})$ is closed under intersection, that is, if $\{B_i \mid i \in I\} \subseteq \Sigma(\mathbf{A})$, then $\bigcap_{i \in I} B_i \in \Sigma(\mathbf{A})$.

Proof. It is obvious that $A \in \Sigma(\mathbf{A})$. Suppose that $\{B_i \mid i \in I\} \subseteq \Sigma(\mathbf{A})$. Let $B = \bigcap_{i \in I} B_i$. If $\mathbf{b} \in \mathscr{D}(F_\xi) \cap B^{\tau(\xi)}$, then $\mathbf{b} \in \mathscr{D}(F_\xi) \cap B_i^{\tau(\xi)}$ for all $i \in I$. Therefore, since $B_i \in \Sigma(\mathbf{A})$, it follows that $F_\xi(\mathbf{b}) \in B_i$. Hence, $F_\xi(\mathbf{b}) \in \bigcap_{i \in I} B_i = B$. Thus, $B \in \Sigma(\mathbf{A})$.

As an application of 4.4, we can define the subalgebra generated by a subset of a partial algebra.

4.5 DEFINITION. Let $\mathbf{A}$ be a partial algebra, and suppose that $X \subseteq A$. Define

$$[X] = \bigcap\{B \in \Sigma(\mathbf{A}) \mid X \subseteq B\}.$$

Then $[X]$ determines a subalgebra of $\mathbf{A}$, which is called the *subalgebra generated by X*.

Strictly speaking $[X]$ depends on $\mathbf{A}$. When it is useful to indicate this dependence we will write $[X]_{\mathbf{A}}$ instead of $[X]$.

In particular, $[\varnothing]$ is the intersection of all sets in $\Sigma(\mathbf{A})$, so that it is the smallest subalgebra of $\mathbf{A}$. If none of the operations F_ξ of $\mathbf{A}$ is zero-ary, then $\varnothing$ determines a subalgebra of $\mathbf{A}$, so that $[\varnothing] = \varnothing$. However, if there is a ξ such that $\tau(\xi) = 0$ and $F_\xi \neq \varnothing$, then every $B \in \Sigma(\mathbf{A})$ contains $F_\xi(0)$. Therefore, $[\varnothing] \neq \varnothing$ in this case.

4.6 EXAMPLES. (a) Let $\mathbf{M} = \langle M; L\rangle$ be the limit algebra of a metric space M. (See Example 1.15.) If X is a subset of M, then $\mathbf{M} \restriction X$ is plainly the limit algebra of the metric space X with the topology inherited from M. However, X determines a subalgebra of $\mathbf{M}$ if and only if X is closed. That is, $\Sigma(\mathbf{M})$ is exactly the set of all closed subspaces of M. Moreover, for an arbitrary set X in M, $[X]$ is the closure of X in M.

(b) Let $\mathbf{F} = \langle F; 0, 1, +, -, \cdot, ^{-1}\rangle$ be a field, considered as a partial algebra. (See Example 1.11.) If $E \in \Sigma(\mathbf{F})$, then $\mathbf{F} \restriction E$ is a field. Indeed, $\mathbf{F} \restriction E$ is clearly a ring containing 1. Moreover, if $x \in E$ and $x \neq 0$, then $x \in \mathscr{D}(^{-1})$. Therefore, since E determines a subalgebra of $\mathbf{F}$, it follows that $x^{-1} \in E$.

There are three basic properties of the operation $X \to [X]$, which are used frequently, usually without special mention:

(a) $X \subseteq Y$ implies $[X] \subseteq [Y]$.
(b) $X \subseteq [X]$.
(c) $[[X]] = [X]$.

That is, $X \to [X]$ satisfies the axioms of a *closure operation*.

4.7 PROPOSITION. Let $\mathbf{A}$ be a partial algebra. Then a subalgebra of a subalgebra of $\mathbf{A}$ is a subalgebra of $\mathbf{A}$. In other words, if $B \in \Sigma(\mathbf{A})$, then $\Sigma(\mathbf{A} \restriction B) = \{C \in \Sigma(\mathbf{A}) \mid C \subseteq B\}$.

Proof. Assume that $\mathbf{A} = \langle A; F_\xi\rangle_{\xi<\rho}$. Then $\mathbf{A} \restriction B = \langle B; F_\xi \cap B^{\tau(\xi)+1}\rangle_{\xi<\rho}$. If $C \in \Sigma(\mathbf{A} \restriction B)$, then clearly $C \subseteq B$. Suppose that $\mathbf{c} \in C^{\tau(\xi)} \cap \mathscr{D}(F_\xi)$. Then

$\mathbf{c} = \mathbf{f} \restriction \tau(\xi)$, where $\mathbf{f} \in F_\xi$. Since B is a subalgebra of $\mathbf{A}$ and $C \subseteq B$, it follows that $\mathbf{f}(\tau(\xi)) \in B$. Thus, $\mathbf{f} \in F_\xi \cap B^{\tau(\xi)+1}$. Therefore, $\mathbf{c} \in C^{\tau(\xi)} \cap \mathscr{D}(F_\xi \cap B^{\tau(\xi)+1})$ Finally, because $C \in \Sigma(\mathbf{A} \restriction B)$, it follows that $F_\xi(\mathbf{c}) = \mathbf{c}(\tau(\xi)) = (F_\xi \cap B^{\tau(\xi)+1})(\mathbf{c}) \in C$. This shows that $C \in \Sigma(\mathbf{A})$. Conversely, if $C \subseteq B$ and $C \in \Sigma(\mathbf{A})$, it is clear that $C \in \Sigma(\mathbf{A} \restriction B)$.

It follows from this result that if $\mathbf{B}$ is a subalgebra of $\mathbf{A}$, and if $X \subseteq B$, then $[X]_\mathbf{B} = [X]_\mathbf{A}$. Indeed, $[X]_\mathbf{B} \in \Sigma(\mathbf{A})$, so that $[X]_\mathbf{A} \subseteq [X]_\mathbf{B} \subseteq B$. Hence, $[X]_\mathbf{A} \in \Sigma(\mathbf{B})$, and $[X]_\mathbf{B} = [X]_\mathbf{A}$.

4.8 Proposition. Let $\mathbf{A}$ and $\mathbf{B}$ be similar partial algebras. Suppose that $\varphi \in \text{Hom}(\mathbf{A}, \mathbf{B})$. Let $C \in \Sigma(\mathbf{B})$. Then $\varphi^{-1}(C) \in \Sigma(\mathbf{A})$.

Proof. Let $D = \varphi^{-1}(C)$. Suppose that $\mathbf{A} = \langle A; F_\xi \rangle_{\xi < \rho}$, $\mathbf{B} = \langle B; G_\xi \rangle_{\xi < \rho}$. Let $\mathbf{a} \in \mathscr{D}(F_\xi) \cap D^{\tau(\xi)}$. Then $\varphi \circ \mathbf{a} \in \mathscr{D}(G_\xi) \cap \varphi(D)^{\tau(\xi)} \subseteq \mathscr{D}(G_\xi) \cap C^{\tau(\xi)}$. Consequently, since φ is a homomorphism and $C \in \Sigma(\mathbf{B})$, $\varphi(F_\xi(\mathbf{a})) = G_\xi(\varphi \circ \mathbf{a}) \in C$. Hence, $F_\xi(\mathbf{a}) \in \varphi^{-1}(C) = D$. This shows that $D \in \Sigma(\mathbf{A})$.

4.9 Corollary. If $\varphi \in \text{Hom}(\mathbf{A}, \mathbf{B})$ where $\mathbf{A}$ and $\mathbf{B}$ are similar partial algebras, then $\varphi([X]) \subseteq [\varphi(X)]$ for all $X \subseteq A$, and $[\varphi^{-1}(Y)] \subseteq \varphi^{-1}([Y])$ for all $Y \subseteq B$. If φ is an isomorphism, then $[\varphi(X)] = \varphi([X])$.

Proof. $X \subseteq \varphi^{-1}([\varphi(X)]) \in \Sigma(\mathbf{A})$ by Proposition 4.8. Therefore by Definition 4.5, $[X] \subseteq \varphi^{-1}([\varphi(X)])$, that is, $\varphi([X]) \subseteq [\varphi(X)]$. From this inclusion, it follows that $\varphi([\varphi^{-1}(Y)]) \subseteq [\varphi(\varphi^{-1}(Y))] \subseteq [Y]$. Hence $[\varphi^{-1}(Y)] \subseteq \varphi^{-1}([Y])$. If φ is an isomorphism, then X can replace Y and φ can replace φ^{-1} in this last inclusion. Hence, $[\varphi(X)] = \varphi([X])$.

4.10 Proposition. Let $\mathbf{A}$ be an abstract algebra, $\mathbf{B}$ a similar partial algebra, and $\varphi \in \text{Hom}(\mathbf{A}, \mathbf{B})$. Then

$$\varphi(\Sigma(\mathbf{A})) \subseteq \Sigma(\mathbf{B}).$$

Proof. Let $\mathbf{A} = \langle A; F_\xi \rangle_{\xi < \rho}$, $\mathbf{B} = \langle B; G_\xi \rangle_{\xi < \rho}$, and $C \in \Sigma(\mathbf{A})$. Let $D = \varphi(C)$. Suppose that $\mathbf{d} \in \mathscr{D}(G_\xi) \cap D^{\tau(\xi)}$. Choose $\mathbf{c} \in C^{\tau(\xi)}$ satisfying $\varphi \circ \mathbf{c} = \mathbf{d}$. Since $\mathbf{A}$ is an algebra, $\mathbf{c} \in \mathscr{D}(F_\xi)$. Moreover, since φ is a homomorphism and $C \in \Sigma(\mathbf{A})$, $G_\xi(\mathbf{d}) = G_\xi(\varphi \circ \mathbf{c}) = \varphi(F_\xi(\mathbf{c})) \in \varphi(C) = D$. Thus, $D \in \Sigma(\mathbf{B})$.

The hypothesis that $\mathbf{A}$ is an algebra cannot be omitted in 4.10, or even replaced by the assumption that φ is an epimorphism (see Problems 5 and 6).

4.11 Corollary. Let $\mathbf{A}$ be an abstract algebra, $\mathbf{B}$ a partial algebra of the same type, and $\varphi \in \text{Hom}(\mathbf{A}, \mathbf{B})$. Suppose that $X \subseteq A$. Then $[\varphi(X)] = \varphi([X])$.

Proof. By 4.10, $\varphi([X]) \in \Sigma(\mathbf{B})$. Clearly, $\varphi([X]) \supseteq \varphi(X)$. Hence, $\varphi([X]) \supseteq [\varphi(X)]$. By 4.9, $\varphi([X]) \subseteq [\varphi(X)]$.

We conclude this section with a lemma that will be needed in the last two chapters.

4.12 LEMMA. Let φ and ψ be homomorphisms of the partial algebra $\mathbf{A} = \langle A; F_\xi \rangle_{\xi < \rho}$ to the partial algebra $\mathbf{B} = \langle B; G_\xi \rangle_{\xi < \rho}$. Suppose that $X \subseteq A$ and $\varphi \restriction X = \psi \restriction X$. Then $\varphi \restriction [X] = \psi \restriction [X]$.

Proof. Let $C = \{a \in A \mid \varphi(a) = \psi(a)\}$. It is sufficient to prove that $C \in \Sigma(\mathbf{A})$. Suppose that $\mathbf{a} \in \mathscr{D}(F_\xi) \cap C^{\tau(\xi)}$. Then $\varphi(F_\xi(\mathbf{a})) = G_\xi(\varphi \circ \mathbf{a}) = G_\xi(\psi \circ \mathbf{a}) = \psi(F_\xi(\mathbf{a}))$. Hence, $F_\xi(\mathbf{a}) \in C$. Therefore, $C \in \Sigma(\mathbf{A})$.

5. Direct Products

The last basic concept that will be studied in this chapter is the notion of a direct product. The direct product of relational systems is defined in terms of Cartesian products of sets. It is convenient to introduce some notation for such products before defining direct products.

5.1 DEFINITION. Let $\{A_i \mid i \in I\}$ be a set of sets. For each $j \in I$, denote by π_j the projection of the Cartesian product $\mathbf{X}_{i\in I} A_i$ into its jth component.

Recall the convention, discussed earlier according to which the Cartesian product $\mathbf{X}_{i\in I} A_i$ is identified with the set of all mappings a of I into $\bigcup_{i\in I} A_i$ such that $a(i) \in A_i$. Adopting this convention, we have the explicit definition

$$\pi_i(a) = a(i).$$

Moreover, from this point of view the following facts are obvious.

5.2 LEMMA. (a) If a and b are elements of $\mathbf{X}_{i\in I} A_i$ such that $\pi_j(a) = \pi_j(b)$ for all $j \in I$, then $a = b$.

(b) If an element $a_i \in A_i$ is given for each $i \in I$, then there is a unique $a \in \mathbf{X}_{i\in I} A_i$ such that $\pi_i(a) = a_i$ for all $i \in I$.

(c) If $A_i \neq \varnothing$ for all $i \in I$, then π_j maps $\mathbf{X}_{i\in I} A_i$ onto A_j.

5.3 DEFINITION. For each i in the index set I, let $\mathbf{A}_i = \langle A_i; R_{i\xi} \rangle_{\xi < \rho}$ be a relational system of type τ (independent of i). The *direct product*# of $\{\mathbf{A}_i \mid i \in I\}$ is the relational system (of type τ)

$$\Pi_{i\in I} \mathbf{A}_i = \langle A; R_\xi \rangle_{\xi < \rho},$$

where $A = \mathbf{X}_{i\in I} A_i$, and R_ξ is the $\tau(\xi)$-ary relation on A defined by

$$R_\xi = \{\mathbf{a} \in A^{\tau(\xi)} \mid \pi_i \circ \mathbf{a} \in R_{i\xi} \text{ for all } i \in I\}.$$

It is not assumed that the systems $\mathbf{A}_i$ in this definition are different. In the extreme case that all $\mathbf{A}_i$ are the same relational system $\mathbf{B}$, we call $\Pi_{i\in I} \mathbf{A}_i$

Unfortunately, many terms other than direct product are commonly used to describe this notion. For example, Birkhoff calls the algebra $\Pi_{i\in I} \mathbf{A}_i$ the *direct union*, while in the theory of Abelian groups it is called the *complete direct sum*. However the term direct product seems to be most frequently used at the present time.

a *direct power* of $\mathbf{B}$, and denote this system by $\mathbf{B}^I$. If $\mathbf{B} = \langle B; S_\xi\rangle_{\xi<\rho}$, then $\mathbf{B}^I = \langle B^I; R_\xi\rangle_{\xi<\rho}$, where R_ξ is the relation on B^I that is associated with S_ξ^I by the natural correspondence $(B^I)^{\tau(\xi)} \to (B^{\tau(\xi)})^I$.

The definition of the direct product of relational systems provides definitions of the direct product of partial algebras and algebras as special cases, provided we identify partial operations with relations in the usual way.

5.4 LEMMA. If $\mathbf{A}_i = \langle A_i; F_{i\xi}\rangle_{\xi<\rho}$ for $i \in I$, where $F_{i\xi}$ is a partial operation for all i, and if $\Pi_{i\in I}\mathbf{A}_i = \langle A; F_\xi\rangle_{\xi<\rho}$, where $A = \mathbf{X}_{i\in I}A_i$, then the relation F_ξ is a partial operation, such that

$$\mathscr{D}(F_\xi) = \mathbf{X}_{i\in I}\mathscr{D}(F_{i\xi}),$$

and for all $i \in I$

$$\pi_i(F_\xi(\mathbf{a})) = F_{i\xi}(\pi_i(\mathbf{a})) \quad \text{if} \quad \mathbf{a} \in \mathscr{D}(F_\xi).$$

Proof. By definition, $\mathbf{f} \in F_\xi$ is equivalent to $\pi_i \circ \mathbf{f} \in F_{i\xi}$ for all i. Since the $F_{i\xi}$ are partial operations, this inclusion is in turn equivalent to the two conditions

(i) $\pi_i \circ (\mathbf{f} \restriction \tau(\xi)) = (\pi_i \circ \mathbf{f}) \restriction \tau(\xi) \in \mathscr{D}(F_{i\xi})$, and
(ii) $\pi_i(\mathbf{f}(\tau(\xi))) = F_{i\xi}(\pi_i \circ (\mathbf{f} \restriction \tau(\xi)))$,

for all $i \in I$. From these observations, the statements of the lemma are easily obtained. In fact, if $\mathbf{f}$ and $\mathbf{g}$ are elements of F_ξ such that $\mathbf{f} \restriction \tau(\xi) = \mathbf{g} \restriction \tau(\xi)$, then for all i, $\pi_i(\mathbf{f}(\tau(\xi))) = F_{i\xi}(\pi_i \circ (\mathbf{f} \restriction \tau(\xi))) = F_{i\xi}(\pi_i \circ (\mathbf{g} \restriction \tau(\xi))) = \pi_i(\mathbf{g}(\tau(\xi)))$. Hence $\mathbf{f}(\tau(\xi)) = \mathbf{g}(\tau(\xi))$ by 5.2. Therefore, F_ξ is a partial operation. Moreover, $\mathbf{a} \in \mathscr{D}(F_\xi)$ if and only if $\mathbf{a} = \mathbf{f} \restriction \tau(\xi)$ for some $\mathbf{f} \in F_\xi$. If this condition is satisfied, then by (i) $\pi_i \circ \mathbf{a} = \pi_i \circ (\mathbf{f} \restriction \tau(\xi)) \in \mathscr{D}(F_{i\xi})$. Therefore, $\mathbf{a} \in \mathbf{X}_{i\in I}\mathscr{D}(F_{i\xi})$. Conversely, suppose that $\mathbf{a} \in \mathbf{X}_{i\in I}\mathscr{D}(F_{i\xi})$. That is, $\pi_i \circ \mathbf{a} \in \mathscr{D}(F_{i\xi})$ for all $i \in I$. Define $\mathbf{f} \in A^{\tau(\xi)+1}$ by the conditions

$$\mathbf{f}(\eta) = \mathbf{a}(\eta) \quad \text{for} \quad \eta < \tau(\xi).$$
$$\pi_i(\mathbf{f}(\tau(\xi))) = F_{i\xi}(\pi_i \circ (\mathbf{f} \restriction \tau(\xi))) \quad \text{for all} \quad i \in I.$$

By 5.2, this last set of equalities uniquely determines $\mathbf{f}(\tau(\xi))$. It is evident that $\mathbf{a} = \mathbf{f} \restriction \tau(\xi)$, so that $\mathbf{f}$ satisfies conditions (i) and (ii) above. Therefore, $\mathbf{f} \in F_\xi$. Hence, $\mathbf{a} \in \mathscr{D}(F_\xi)$. This shows that $\mathscr{D}(F_\xi) = \mathbf{X}_{i\in I}\mathscr{D}(F_{i\xi})$. Finally, suppose that $\mathbf{a} \in \mathscr{D}(F_\xi)$. Let $\mathbf{a} = \mathbf{f} \restriction \tau(\xi)$, where $\mathbf{f} \in F_\xi$. Then by (ii), $\pi_i(F_\xi(\mathbf{a})) = \pi_i(\mathbf{f}(\tau(\xi))) = F_{i\xi}(\pi_i \circ (\mathbf{f} \restriction \tau(\xi))) = F_{i\xi}(\pi_i \circ \mathbf{a})$ for all $i \in I$. The proof is therefore complete.

As a corollary of 5.4, it follows that the direct product of partial algebras is a partial algebra, and the direct product of algebras is an algebra.

Comment. The direct product of the empty set of relational systems is worthy of special consideration. By definition, this product is the system**

$$\mathbf{1} = \langle 1; 1^{\tau(\xi)}\rangle_{\xi<\rho}.$$

** Recall that $1 = \{\varnothing\}$.

Thus, the empty product is uniquely determined by the type τ. Note that **1** is always an abstract algebra.

It is readily seen from Definition 5.3 that the projection mappings $\pi_i:\Pi_{i\in I}\mathbf{A}_i \to \mathbf{A}_i$ are homomorphisms. The following extension of 5.2(b) shows that generally these mappings are epimorphisms.

5.5 LEMMA. Let $\mathbf{A}_i = \langle A_i;\, R_{i\xi}\rangle_{\xi<\rho}$ be relational systems of type $\tau \in \mathbf{Ord}^\rho$. Define $\mathbf{A} = \langle A;\, R_\xi\rangle_{\xi<\rho} = \Pi_{i\in I}\mathbf{A}_i$. Let $\xi < \rho$. If $\mathbf{a}_i \in R_{i\xi}$ is given for each $i \in I$, then $\mathbf{a} \in R_\xi$ exists satisfying $\pi_i \circ \mathbf{a} = \mathbf{a}_i$ for all $i \in I$. Consequently, if $R_{i\xi} \neq \varnothing$ for all $i \in I$, then each π_i is an epimorphism.

Proof. Since $\mathbf{a}_i(\eta) \in A_i$ for each $\eta < \tau(\xi)$, it follows from 5.2(b) that there exists $\mathbf{a}(\eta) \in A$ satisfying $\pi_i(\mathbf{a}(\eta)) = \mathbf{a}_i(\eta)$ for all $i \in I$. Consequently, $\pi_i \circ \mathbf{a} = \mathbf{a}_i$ for all i, and therefore $\mathbf{a} \in R_\xi$ by Definition 5.3. The last statement†† follows from this observation and 5.2(c).

The most important property of direct products is given in the next theorem.

5.6 THEOREM. Let $\{A_i \mid i \in I\}$ be a set of relational systems of similarity type $\tau \in \mathrm{Ord}^\rho$. Suppose that $\mathbf{B}$ is a relational system of type τ. Assume that for each $i \in I$ a homomorphism $\varphi_i:\mathbf{B} \to \mathbf{A}_i$ is given. Then there is a unique homomorphism $\varphi:\mathbf{B} \to \Pi_{i\in I}\mathbf{A}_i$ such that $\varphi_i = \pi_i \circ \varphi$ for all $i \in I$. If $\mathbf{B}$ and all of the $\mathbf{A}_i$ are partial algebras, then

$$\Gamma_\varphi = \bigcap_{i\in I}\Gamma_{\varphi_i}$$

Proof. Let $\mathbf{B} = \langle B;\, S_\xi\rangle_{\xi<\rho}$, $\mathbf{A}_i = \langle A_i;\, R_{i\xi}\rangle_{\xi<\rho}$, and denote $\Pi_{i\in I}\mathbf{A}_i = \langle A;\, R_\xi\rangle_{\xi<\rho}$. For $b \in B$, define an element $\varphi(b) \in \mathbf{X}_{i\in I}A_i$ by the conditions

(i) $$\pi_i(\varphi(b)) = \varphi_i(b) \quad \text{for all} \quad i \in I.$$

By 5.2, these conditions determine a unique $\varphi(b)$ in $\mathbf{X}_{i\in I}A_i$. Moreover,

$$\pi_i \circ (\varphi \circ S_\xi) = \varphi_i \circ S_\xi \subseteq R_{i\xi}$$

for all $i \in I$. Consequently, $\varphi \circ S_\xi \subseteq R_\xi$. Since ξ was arbitrary, φ is a homomorphism of $\mathbf{B}$ to $\Pi_{i\in I}\mathbf{A}_i$. By (i) above, $\pi_i \circ \varphi = \varphi_i$. As we noted, this condition uniquely determines φ. Assume now that $\mathbf{B}$ and all $\mathbf{A}_i$ are partial algebras. By definition, $\langle a, b\rangle \in \Gamma_\varphi$ is equivalent to $\varphi(a) = \varphi(b)$. By (i) and by (5.2), $\varphi(a) = \varphi(b)$ is equivalent to $\varphi_i(a) = \varphi_i(b)$ for all $i \in I$, that is, $\langle a, b\rangle \in \Gamma_{\varphi_i}$ for all $i \in I$. Hence, $\Gamma_\varphi = \bigcap_{i\in I}\Gamma_{\varphi_i}$.

5.7 COROLLARY. Let I be an index set. Suppose that for each $i \in I$, $\mathbf{B}_i$ and $\mathbf{A}_i$ are similar relational systems, and $\psi_i \in \mathrm{Hom}\,(\mathbf{B}_i, \mathbf{A}_i)$. Then there is a

†† Sometimes in the study of relational systems it is convenient to consider the underlying set as a unary relation of the system. From this point of view, the last statement of this proof is unnecessary.

unique homomorphism ψ of $\Pi_{i\in I}\mathbf{B}_i$ to $\Pi_{i\in I}\mathbf{A}_i$ such that

$$\begin{array}{ccc} \Pi_{i\in I}\mathbf{B}_i & \xrightarrow{\quad\psi\quad} & \Pi_{i\in I}\mathbf{A}_i \\ \downarrow{\scriptstyle \pi_j} & & \downarrow{\scriptstyle \pi_j} \\ \mathbf{B}_j & \xrightarrow[\quad\psi_j\quad]{} & \mathbf{A}_j \end{array}$$

is consistent for all $j \in I$. If each ψ_i is a monomorphism, then ψ is a monomorphism. If each ψ_i is an epimorphism, then ψ is an epimorphism.

Proof. The first statement follows from 5.6 by taking $\mathbf{B} = \Pi_{i\in I}\mathbf{B}_i$ and $\varphi_i = \psi_i \circ \pi_i$. If each ψ_i is a monomorphism and $\psi(b) = \psi(c)$, then $\psi_i(\pi_i(b)) = \psi_i(\pi_i(c))$, so that $\pi_i(b) = \pi_i(c)$ for all $i \in I$. Therefore, $b = c$ by 5.2. Consequently, ψ is a monomorphism. Assume that each ψ is an epimorphism. We can suppose that $B_i \neq \varnothing$ for all $i \in I$. Let $\mathbf{A}_i = \langle A_i; R_{i\xi}\rangle_{\xi<\rho}$, $\Pi_{i\in I}\mathbf{A}_i = \langle A; R_\xi\rangle_{\xi<\rho}$, $\mathbf{B}_i = \langle B_i; S_{i\xi}\rangle_{\xi<\rho}$, and $\Pi_{i\in I}\mathbf{B}_i = \langle B; S_\xi\rangle_{\xi<\rho}$. If $\mathbf{a} \in R_\xi$, then $\pi_i \circ \mathbf{a} \in R_{i\xi}$ for all $i \in I$ by Definition 5.3. Since ψ_i is an epimorphism, there exists $\mathbf{b}_i \in S_{i\xi}$ such that $\psi_i \circ \mathbf{b}_i = \pi_i \circ \mathbf{a}_i$. By 5.5, there is an element $\mathbf{b} \in S_\xi$ such that $\pi_i \circ \mathbf{b} = \mathbf{b}_i$ for all $i \in I$. It follows that $\pi_i \circ (\psi \circ \mathbf{b}) = (\psi_i \circ \pi_i) \circ \mathbf{b} = \psi_i \circ \mathbf{b}_i = \pi_i \circ \mathbf{a}$ for every $i \in I$. Hence, by 5.2, $\psi \circ \mathbf{b} = \mathbf{a}$. A similar argument shows that ψ maps B onto A. Hence ψ is an epimorphism.

In case $\mathbf{B}_i$ is a subalgebra of $\mathbf{A}_i$ and $\psi_i = I_{B_i}$, the second conclusion of 5.7 can be strengthened.

5.8 Lemma. For each $i \in I$, let $\mathbf{B}_i$ be a subalgebra of the partial algebra $\mathbf{A}_i$. Then $\mathbf{X}_{i\in I}B_i \in \Sigma(\Pi_{i\in I}\mathbf{A}_i)$, and $\Pi_{i\in I}\mathbf{B}_i = (\Pi_{i\in I}\mathbf{A}_i) \restriction \mathbf{X}_{i\in I}B_i$.

This result is a direct consequence of 5.4.

5.9 Example. Suppose that for each $i \in I$, $\mathbf{M}_i = \langle M_i; L\rangle$ is the limit algebra of the metric space M_i. Let

$$\mathbf{M} = \Pi_{i\in I}\mathbf{M}_i = \langle M; L\rangle$$

The set M is the cartesian product $\mathbf{X}_{i\in I}M_i$. It is well known that if $|I| \leq \aleph_0$, then a metric can be defined on M such that sequences in M converge if and only if all of their components converge. This gives the usual product topology. Moreover, with such a metric, the limit algebra of M is exactly $\mathbf{M}$. On the other hand, if I is uncountable, then $\langle M; L\rangle$ will generally not be the limit algebra of a metric space.

The case of countable direct products is interesting. Suppose that M is a separable metric space with the metric μ. Let $\{p_n \mid n < \omega\}$ be a dense set of points in M. As usual, let $\mathbf{M} = \langle M; L\rangle$ be the limit algebra of M. For all $n < \omega$, let $\mathbf{M}_n = \langle E; L\rangle$, where E is the closed unit interval [0, 1] with the usual distance metric and L is the associated limit (partial) operation. Define $\varphi_n: M \to E$ by the condition

$$\varphi_n(a) = \frac{\mu(a, p_n)}{1 + \mu(a, p_n)}.$$

It is clear that if $\mathbf{a}(0), \mathbf{a}(1), \mathbf{a}(2), \cdots$ converges to b, then

$$\frac{\mu(\mathbf{a}(0), p_n)}{1 + \mu(\mathbf{a}(0), p_n)}, \frac{\mu(\mathbf{a}(1), p_n)}{1 + \mu(\mathbf{a}(1), p_n)}, \frac{\mu(\mathbf{a}(2), p_n)}{1 + \mu(\mathbf{a}(2), p_n)}, \cdots$$

converges to

$$\frac{\mu(b, p_n)}{1 + \mu(b, p_n)}.$$

In other words, $\varphi_n(L(\mathbf{a})) = L(\varphi_n \circ \mathbf{a})$. Thus, φ_n is a homomorphism. By 5.6, there is a homomorphism φ of $\mathbf{M}$ to $\Pi_{n<\omega}\mathbf{M}_n$ such that $\varphi_n = \pi_n \circ \varphi$ for all $n < \omega$. In topological terms, this means that there is a continuous mapping of M into the Hilbert cube. If M is not compact, then $\varphi(M)$ cannot be closed, so that $\varphi(M) \notin \Sigma(\Pi_{n<\omega}\mathbf{M}_n)$ in this case. However, φ is a monomorphism. Indeed, by 5.6, $\varphi(a) = \varphi(b)$ if and only if $\varphi_n(a) = \varphi_n(b)$ for all $n < \omega$. Suppose that $a \neq b$. Let $\mu(a, b) = r$. Choose n so that $\mu(a, p_n) < r/2$. Then $\mu(b, p_n) > r/2$. Hence,

$$\varphi_n(a) = \frac{\mu(a, p_n)}{1 + \mu(a, p_n)} < \frac{r}{2 + r} \quad \text{and} \quad \varphi_n(b) = \frac{\mu(b, p_n)}{1 + \mu(b, p_n)} > \frac{r}{2 + r}.$$

It follows that $\varphi(a) \neq \varphi(b)$. Finally, it can be shown that φ is an isomorphism of $\mathbf{M}$ onto $(\Pi_{n<\omega}\mathbf{M}_n) \restriction \varphi(M)$. In fact, suppose that $\mathbf{a} \in M^\omega$, $b \in M$, and $\varphi \circ \mathbf{a}$ converges to $\varphi(b)$ in the space $\mathbf{X}_{n<\omega}M_n$, that is, $\varphi_n \circ \mathbf{a}$ converges to $\varphi_n(b)$ for all $n < \omega$. Suppose that $\mathbf{a}$ does not converge to b. Then there exists $r > 0$, and an infinite sequence $n_1 < n_2 < n_3 < \cdots$ such that $\mu(\mathbf{a}(n_i), b) > r$ for all i. Choose n so that $\mu(b, p_n) < r/3$. Then $\mu(\mathbf{a}(n_i), p_n) > 2r/3$. Thus, for all i, $\varphi_n(\mathbf{a}(n_i)) > r/(\frac{3}{2} + r)$ and $\varphi_n(b) < r/(3 + r)$. Therefore, $\varphi_n \circ \mathbf{a}$ does not converge to $\varphi_n(b)$. This contradiction proves that φ is an epimorphism, (and therefore an isomorphism) to $(\Pi_{n<\omega}\mathbf{M}_n) \restriction \varphi(M)$.

In topological terms we have shown the well-known theorem of Urysohn, which states that every separable metric space is homeomorphic to a subspace of the Hilbert cube.

6. Closure Properties of Classes of Algebras

The theory of abstract algebras can be divided into two parts—the theory of a single algebra and the theory of classes of algebras of a given type.

However, even in the study of a single algebra, it is frequently convenient to use some of the elementary ideas that occur in the theory of classes of algebras. The final section of this chapter will be devoted to the introduction of a few of these elementary ideas.

6.1 DEFINITION. Let $\mathfrak{A}$ be a class of partial algebras of type τ. Define:

$\mathscr{I}\mathfrak{A}$ = class of all partial algebras of type τ that are isomorphic to some algebra in $\mathfrak{A}$.

$\mathscr{Q}\mathfrak{A}$ = class of all partial algebras of type τ that are epimorphic images of some partial algebra in $\mathfrak{A}$, that is, $\mathbf{B} \in \mathscr{Q}\mathfrak{A}$ if and only if there is an $\mathbf{A} \in \mathfrak{A}$ and an epimorphism φ of $\mathbf{A}$ to $\mathbf{B}$.

$\mathscr{S}\mathfrak{A}$ = class of all partial algebras of type τ that are isomorphic to a subalgebra of some partial algebra in $\mathfrak{A}$.

$\mathscr{P}\mathfrak{A}$ = class of all partial algebras of type τ that are isomorphic to a direct product of some set of partial algebras in $\mathfrak{A}$.

A class $\mathfrak{A}$ of similar partial algebras is called *abstract* if $\mathscr{I}\mathfrak{A} = \mathfrak{A}$. Generally speaking modern algebra is concerned only with properties of partial algebras which are invariant under isomorphism. Thus, attention is focused mainly on abstract classes. It should be noted that the definitions of the operators $\mathscr{Q}$, $\mathscr{S}$, and $\mathscr{P}$ are formulated so that $\mathscr{Q}\mathfrak{A}$, $\mathscr{S}\mathfrak{A}$, and $\mathscr{P}\mathfrak{A}$ are abstract classes, no matter what $\mathfrak{A}$ might be. In fact, $\mathscr{I}\mathscr{Q}\mathfrak{A} = \mathscr{Q}\mathscr{I}\mathfrak{A} = \mathscr{Q}\mathfrak{A}$, $\mathscr{I}\mathscr{S}\mathfrak{A} = \mathscr{S}\mathscr{I}\mathfrak{A} = \mathscr{S}\mathfrak{A}$, and $\mathscr{I}\mathscr{P}\mathfrak{A} = \mathscr{P}\mathscr{I}\mathfrak{A} = \mathscr{P}\mathfrak{A}$.

By the comment following 5.4, $\mathscr{P}\mathfrak{A}$ contains all one-element partial algebras of type τ. Indeed, any such algebra is isomorphic to the product of the empty set of algebras. This remark will be used in Chapters 4 and 5.

There are some elementary properties of the operators $\mathscr{I}$, $\mathscr{Q}$, $\mathscr{S}$, and $\mathscr{P}$ which are used very often. We present these properties in two groups.

6.2 *Proposition.* The operators $\mathscr{I}$, $\mathscr{Q}$, $\mathscr{S}$, and $\mathscr{P}$ are closure operators. That is, if $\mathscr{C}$ is one of $\mathscr{I}$, $\mathscr{Q}$, $\mathscr{S}$, or $\mathscr{P}$, then for any classes $\mathfrak{A}$ and $\mathfrak{B}$ of partial algebras of type τ,

(a) $\mathfrak{A} \subseteq \mathfrak{B}$ implies $\mathscr{C}\mathfrak{A} \subseteq \mathscr{C}\mathfrak{B}$.
(b) $\mathfrak{A} \subseteq \mathscr{C}\mathfrak{A}$.
(c) $\mathscr{C}\mathscr{C}\mathfrak{A} = \mathscr{C}\mathfrak{A}$.

Proof. The property (a) is plainly satisfied for $\mathscr{C} = \mathscr{I}$, $\mathscr{Q}$, $\mathscr{S}$, or $\mathscr{P}$. Since I_A is an epimorphism of $\mathbf{A}$ to $\mathbf{A}$, it follows that $\mathfrak{A} \subseteq \mathscr{I}\mathfrak{A}$ and $\mathfrak{A} \subseteq \mathscr{Q}\mathfrak{A}$. The inclusion $\mathfrak{A} \subseteq \mathscr{S}\mathfrak{A}$ is a consequence of the observation that $A \in \Sigma(\mathbf{A})$ (see 4.4). It is clear that $\mathbf{A}^1 \cong \mathbf{A}$, so that $\mathfrak{A} \subseteq \mathscr{P}\mathfrak{A}$. The equalities in (c) for $\mathscr{C} = \mathscr{I}$, $\mathscr{Q}$, and $\mathscr{S}$ are consequences of 2.5 and 4.7. In order to show that $\mathscr{P}\mathscr{P}\mathfrak{A} = \mathscr{P}\mathfrak{A}$, it is sufficient to prove

$$\Pi_{i\in I}(\Pi_{j\in J}\mathbf{A}_{ij}) \cong \Pi_{\langle i,j\rangle\in I\times J}\mathbf{A}_{ij}.$$

This isomorphism is obtained from the natural correspondence between $(A^J)^I$ and $A^{I\times J}$ (where $A = \bigcup_{i\in I}\bigcup_{j\in J}A_{ij}$). We omit the details.

6.3 PROPOSITION. Let $\mathfrak{A}$ be any class of algebras of type τ. Then

(a) $\mathscr{S}\mathscr{Q}\mathfrak{A} \subseteq \mathscr{Q}\mathscr{S}\mathfrak{A}$.
(b) $\mathscr{P}\mathscr{S}\mathfrak{A} \subseteq \mathscr{S}\mathscr{P}\mathfrak{A}$.
(c) $\mathscr{P}\mathscr{Q}\mathfrak{A} \subseteq \mathscr{Q}\mathscr{P}\mathfrak{A}$.

Proof. (a) Let $\mathbf{A} \in \mathfrak{A}$, and $\varphi:\mathbf{A} \to \mathbf{B}$ be an epimorphism. Suppose that $C \in \Sigma(\mathbf{B})$. By 4.8, $D = \varphi^{-1}(C) \in \Sigma(\mathbf{A})$. We prove that $\varphi \restriction D$ is an epimorphism of $\mathbf{A} \restriction D$ to $\mathbf{B} \restriction C$, which shows that $\mathbf{B} \restriction C \in \mathscr{Q}\mathscr{S}\mathfrak{A}$. Let $\mathbf{A} = \langle A; F_\xi \rangle_{\xi < \rho}$ and $\mathbf{B} = \langle B; G_\xi \rangle_{\xi < \rho}$. Suppose that $\mathbf{g} \in G_\xi \cap C^{\tau(\xi)+1}$. Since φ is an epimorphism, $\mathbf{f} \in F_\xi$ exists satisfying $\varphi \circ \mathbf{f} = \mathbf{g}$. Therefore, by definition, $\mathbf{f} \in D^{\tau(\xi)+1}$. Consequently, $\varphi \circ (F_\xi \cap D^{\tau(\xi)+1}) = G_\xi \cap C^{\tau(\xi)+1}$. The inclusion (b) is a restatement of the last parts of 5.8, while (c) is similarly related to 5.7.

6.4 COROLLARY. Let $\mathfrak{A}$ be a class of algebras of type τ. Define $\hat{\mathfrak{A}} = \mathscr{Q}\mathscr{S}\mathscr{P}\mathfrak{A}$. Then $\mathscr{Q}\hat{\mathfrak{A}} = \hat{\mathfrak{A}}$, $\mathscr{S}\hat{\mathfrak{A}} = \hat{\mathfrak{A}}$, and $\mathscr{P}\hat{\mathfrak{A}} = \hat{\mathfrak{A}}$.

The classes such as the $\hat{\mathfrak{A}}$ in 6.4 which are closed under the operators $\mathscr{Q}$, $\mathscr{S}$, and $\mathscr{P}$ will be examined more completely in the last chapter.

Problems

1. Let $\mathbf{A} = \langle A; F_\xi \rangle_{\xi < \rho}$ be a partial algebra of type τ. Prove that $\Sigma(\mathbf{A})$ contains all subsets of A if and only if $F_\xi(\mathbf{a}) \in \{\mathbf{a}(\eta) \mid \eta < \tau(\xi)\}$ for every $\xi < \rho$ and $\mathbf{a} \in \mathscr{D}(F_\xi)$.

2. Let $\mathbf{A} = \langle A; F_\xi \rangle_{\xi < \rho}$ be an algebra of type $\tau \in \omega^\rho$ (so that each F_ξ is a finitary operation). Prove that $\Theta(\mathbf{A}) = E(A)$ if and only if either (i) $|A| \leq 2$, or (ii) each F_ξ is a constant operation (that is, $F_\xi(\mathbf{a}) = F_\xi(\mathbf{b})$ for all $\mathbf{a}$, $\mathbf{b}$ in $A^{\tau(\xi)}$) or a projection operation (that is, there is an $i < \tau(\xi)$ such that $F_\xi(\mathbf{a}) = \mathbf{a}(i)$ for all $\mathbf{a} \in A^{\tau(\xi)}$).

3. Let $\mathbf{A} = \langle A; R_\xi \rangle_{\xi < \rho}$ be a relational system of type τ. Define $R_\xi^+ : R_\xi \to A$ by $R_\xi^+(\mathbf{a}) = \mathbf{a}(0)$ for $\mathbf{a} \in R_\xi$, $\mathbf{A}^+ = \langle A; R_\xi^+ \rangle_{\xi < \rho}$.

(a) Prove that $\mathbf{A}^+$ is a partial algebra of type τ.
(b) Prove that $\Theta(\mathbf{A}^+) = E(A)$ and $\Sigma(\mathbf{A}^+) = \mathbf{P}(A)$.
(c) Prove that Hom $(\mathbf{A}^+, \mathbf{B}^+) =$ Hom $(\mathbf{A}, \mathbf{B})$.

These results can be interpreted as evidence that partial algebras are no less general than relational systems.

4. Let $\mathbf{S}$ be a semigroup with identity. Prove that there exists an abstract algebra $\mathbf{A}$ such that Hom $(\mathbf{A}, \mathbf{A})$ (with the binary operation of composition) is isomorphic to $\mathbf{S}$. Show that any group is isomorphic to the group of all automorphisms of a suitable abstract algebra.

5. Let R be the real line with the metric space topology. Let C be the circle in the complex plane with the metric topology. Show that the mapping $\varphi: R \to C$ given by $\varphi(x) = e^{2\pi ix}$ is an epimorphism of the limit algebra $\mathbf{M}_1 = \langle R; L\rangle$ to the limit algebra $\mathbf{M}_2 = \langle C; L\rangle$, but that $\varphi(\Sigma(\mathbf{M}_1)) \nsubseteq \Sigma(\mathbf{M}_2)$.

6. Let $\mathbf{B} = \langle B; G_\xi\rangle_{\xi<\rho}$ be a partial algebra of type $\tau \in \mathrm{Ord}^\rho$. Suppose that X is a nonempty set, which is disjoint from B and $\varphi \in B^X$ is such that $\varphi(X) \notin \Sigma(\mathbf{B})$. Define $\mathbf{A} = \langle B \cup X; G_\xi\rangle_{\xi<\rho}$. Prove that $\mathbf{A}$ is a partial algebra of type τ and $\psi = I_B \cup \varphi$ is an epimorphism of $\mathbf{A}$ to $\mathbf{B}$. Show that if $\tau(\xi) > 0$ for all $\xi < \rho$, then $X \in \Sigma(\mathbf{A})$; deduce that in this case $\psi(\Sigma(\mathbf{A})) \nsubseteq \Sigma(\mathbf{B})$.

7. Prove that if $\mathbf{A} = \langle A; F_\xi\rangle_{\xi<\rho}$ is a partial algebra, then there is an algebra $\mathbf{B} = \langle B; G_\xi\rangle_{\xi<\rho}$ such that $A \subseteq B$ and $\mathbf{A} = \mathbf{B} \upharpoonright A$.

8. Let $\mathbf{A} = \langle A; F_\xi\rangle_{\xi<\rho}$ be a partial algebra of type τ. Assume for convenience that $A \cap \rho = \varnothing$. Let i and j be distinct indices that are not in $A \cup \rho$. Set $K = \rho \cup A \cup \{i, j\}$. Define a partial algebra of type $\sigma = \rho \cup \{\langle a, 0\rangle \mid a \in A\} \cup \{\langle i, 1\rangle, \langle j, 2\rangle\}$ as follows:

$$\mathbf{C}(\mathbf{A}) = \langle A \times A; G_\xi, G_a, G_i, G_j\rangle_{\xi<\rho, a\in A}$$

where $\langle A \times A; G_\xi\rangle_{\xi<\rho}$ is the direct product of $\mathbf{A}$ with itself, G_a is the zero-ary operation $\langle 0, \langle a, a\rangle\rangle$ for each $a \in A$, $G_i = \{\langle\langle a, b\rangle, \langle b, a\rangle\rangle \mid a, b \in A\}$, and $G_j = \{\langle\langle a, b\rangle, \langle b, c\rangle, \langle a, c\rangle\rangle \mid a, b, c \in A\}$. The partial algebra $\mathbf{C}(\mathbf{A})$ is called the *congruence algebra* of A. Prove that $\Sigma(\mathbf{C}(\mathbf{A})) = \Theta(\mathbf{A})$.

9. Let $J = \bigcup_{i\in I} J_i$. Suppose that $\mathbf{A} = \langle A; F_j\rangle_{j\in J}$ is a partial algebra whose type is indexed by J. Define $\mathbf{A}_i = \langle A; F_j\rangle_{j\in J_i}$. Prove that $\Theta(\mathbf{A}) = \bigcap_{i\in I}\Theta(\mathbf{A}_i)$ and $\Sigma(\mathbf{A}) = \bigcap_{i\in I}\Sigma(\mathbf{A}_i)$.

10. Let $\{\mathbf{A}_i \mid i \in I\}$ be a set of relational systems of type τ. For $U \subseteq I$, define the mapping $\pi_{U,I}: \mathbf{X}_{i\in I} A_i \to \mathbf{X}_{i\in U} A_i$ by $\pi_{U,I}(a) = a \upharpoonright U$.

(a) Prove that $\pi_{U,I}$ is an epimorphism of $\Pi_{i\in I}\mathbf{A}_i$ to $\Pi_{i\in U}\mathbf{A}_i$.

(b) Show that if $V \subseteq U \subseteq I$, then $\pi_{V,I} = \pi_{V,U} \circ \pi_{U,I}$.

(c) Let $\mathbf{B}$ be a relational system of type τ. Suppose that for each $i \in I$, $\varphi_i \in \mathrm{Hom}\,(\mathbf{B}, \mathbf{A}_i)$. Let $\varphi \in \mathrm{Hom}\,(\mathbf{B}, \Pi_{i\in I}\mathbf{A}_i)$ satisfy $\varphi_i = \pi_i \circ \varphi$ for all $i \in I$, and $\psi \in \mathrm{Hom}\,(\mathbf{B}, \Pi_{i\in U}\mathbf{A}_i)$ satisfy $\varphi_i = \pi_i \circ \psi$ for all $i \in U$. Prove that $\psi = \pi_{U,I} \circ \varphi$.

11. Let $\mathfrak{A}$ be a class of partial algebras of type τ. Define $\mathscr{R}\mathfrak{A}$ to be the class of all partial algebras of type τ such that $\bigcap\{\Gamma \in \Theta(\mathbf{A}) \mid \mathbf{A}/\Gamma \in \mathscr{I}\mathfrak{A}\} = I_A$. Prove the following results.

(a) $\mathscr{R}$ is a closure operator.

(b) $\mathscr{R}\mathfrak{A}$ is an abstract class.

(c) $\mathscr{P}\mathfrak{A} \subseteq \mathscr{R}\mathfrak{A}$.

(d) If $\mathfrak{A}$ is a class of algebras, then $\mathscr{S}\mathscr{R}\mathfrak{A} \subseteq \mathscr{R}\mathscr{S}\mathfrak{A}$, and $\mathscr{R}\mathscr{Q}\mathfrak{A} \subseteq \mathscr{Q}\mathscr{R}\mathfrak{A}$.

Remark. A class $\mathfrak{A}$ of similar partial algebras is called *residual* if $\mathscr{R}\mathfrak{A} = \mathfrak{A}$.

12. Let $\mathfrak{A}$ be a class of partial algebras of type τ. For any partial algebra $\mathbf{A}$ of type τ, define

$$N_{\mathfrak{A}}(\mathbf{A}) = \bigcap\{\Gamma \in \Theta(\mathbf{A}) \mid \mathbf{A}/\Gamma \in \mathscr{I}\mathfrak{A}\}.$$

Thus, $N_{\mathfrak{A}}(\mathbf{A}) = I_A$ if and only if $\mathbf{A} \in \mathscr{R}\mathfrak{A}$.

(a) Prove that the mapping $N: \mathbf{A} \to N_{\mathfrak{A}}(\mathbf{A})$ satisfies (i) $N(\mathbf{A}/N(\mathbf{A})) = I_{\mathbf{A}/N(\mathbf{A})}$; and (ii) if $\varphi: \mathbf{A} \to \mathbf{B}$ is an epimorphism, then $\varphi \circ N(\mathbf{A}) \subseteq N(\mathbf{B})$.

(b) Show that $N_{\mathfrak{A}} = N_{\mathscr{R}\mathfrak{A}}$.

(c) Prove that if $\mathfrak{A}$ and $\mathfrak{B}$ are residual classes of similar partial algebras such that $N_{\mathfrak{A}} = N_{\mathfrak{B}}$, then $\mathfrak{A} = \mathfrak{B}$.

(d) Let $\mathfrak{B}$ be a class of similar partial algebras such that $\mathscr{Q}\mathfrak{B} = \mathfrak{B}$. Suppose N is a mapping that assigns to each $\mathbf{A} \in \mathfrak{B}$ a congruence relation $N(\mathbf{A}) \in \Theta(\mathbf{A})$ so that the conditions (i) and (ii) of **(a)** are satisfied. Let $\mathfrak{A} = \{\mathbf{A} \in \mathfrak{B} \mid N(\mathbf{A}) = I_A\}$. Prove that $N = N_{\mathfrak{A}}$.

Comment. Most of the familiar "radicals" used in algebra are of the form $N_{\mathfrak{A}}$ for a suitable class $\mathfrak{A}$. In view of **(d)**, this fact is not surprising.

13. Let α be a cardinal number, and suppose that $\mathfrak{A}$ is a class of partial algebras of type τ. Define $\mathscr{S}_\alpha\mathfrak{A}$ to be the class of all partial algebras $\mathbf{B}$ that are isomorphic to some partial algebra of the form $\mathbf{A} \restriction [X]$, where $\mathbf{A} \in \mathfrak{A}$, $X \subseteq A$, and $|X| < \alpha$. It is evident that this definition is phrased so that $\mathscr{S}_\alpha\mathfrak{A}$ is an abstract class. Prove the following assertions (in which $\mathfrak{A}$ and $\mathfrak{B}$ designate classes of similar partial algebras).

(a) $\mathfrak{A} \subseteq \mathfrak{B}$ implies $\mathscr{S}_\alpha\mathfrak{A} \subseteq \mathscr{S}_\alpha\mathfrak{B}$.
(b) $\mathscr{S}_\alpha\mathscr{S}_\alpha\mathfrak{A} = \mathscr{S}_\alpha\mathfrak{A}$.
(c) $\mathscr{S}_\alpha\mathscr{S}\mathfrak{A} = \mathscr{S}_\alpha\mathfrak{A}$.
(d) $\mathscr{S}_\alpha\mathscr{P}\mathfrak{A} = \mathscr{S}_\alpha\mathscr{P}\mathscr{S}_\alpha\mathfrak{A}$ (see 5.8).

Now suppose that $\mathfrak{A}$ is a class of similar algebras. Prove

(e) $\mathscr{S}_\alpha\mathscr{Q}\mathfrak{A} \subseteq \mathscr{Q}\mathscr{S}_\alpha\mathfrak{A}$.

14. As in the previous problem, let α be any cardinal number, and suppose that $\mathfrak{A}$ is a class of partial algebras of type τ. Define $\mathscr{S}_\alpha^{-1}\mathfrak{A}$ to be the class of all partial algebras $\mathbf{A}$ of type τ such that $\mathscr{S}_\alpha\{\mathbf{A}\} \subseteq \mathscr{I}\mathfrak{A}$. Prove the following assertions (in which $\mathfrak{A}$ and $\mathfrak{B}$ designate classes of similar partial algebras).

(a) $\mathfrak{B} \subseteq \mathscr{S}_\alpha^{-1}\mathfrak{A}$ if and only if $\mathscr{S}_\alpha\mathfrak{B} \subseteq \mathscr{I}\mathfrak{A}$.
(b) $\mathscr{S}_\alpha^{-1}\mathscr{S}_\alpha\mathfrak{A} = \mathscr{S}_\alpha^{-1}\mathfrak{A}$.
(c) $\mathscr{S}_\alpha\mathscr{S}_\alpha^{-1}\mathfrak{A} \subseteq \mathscr{S}_\alpha\mathfrak{A} \cap \mathscr{I}\mathfrak{A}$.
(d) $\mathfrak{A} \subseteq \mathfrak{B}$ implies $\mathscr{S}_\alpha^{-1}\mathfrak{A} \subseteq \mathscr{S}_\alpha^{-1}\mathfrak{B}$.
(e) $\mathscr{S}_\alpha\mathscr{S}_\alpha^{-1}\mathfrak{A} \subseteq \mathscr{S}_\alpha^{-1}\mathfrak{A}$.
(f) $\mathscr{S}_\alpha^{-1}\mathscr{S}_\alpha^{-1}\mathfrak{A} = \mathscr{S}_\alpha^{-1}\mathfrak{A}$.
(g) $\mathscr{S}_\alpha\mathfrak{A} \subseteq \mathscr{I}\mathfrak{A}$ implies $\mathfrak{A} \subseteq \mathscr{S}_\alpha^{-1}\mathfrak{A}$.
(h) $\mathscr{S}\mathscr{S}_\alpha^{-1}\mathfrak{A} = \mathscr{S}_\alpha^{-1}\mathfrak{A}$.
(i) $\mathscr{P}\mathscr{S}_\alpha^{-1}\mathfrak{A} \subseteq \mathscr{S}_\alpha^{-1}\mathscr{P}\mathfrak{A}$.

In the remaining statements **(j)** and **(k)**, assume that $\mathfrak{A}$ is a class of *algebras* of type $\tau \in \mathrm{Ord}^j$, and α is a cardinal number which satisfies $\alpha > |\tau(\xi)|$ for all $\xi < \rho$.

(j) If $\mathbf{A} \in \mathscr{S}_\alpha^{-1}\mathfrak{A}$, then $\mathbf{A}$ is an algebra.
(k) $\mathscr{Q}\mathscr{S}_\alpha^{-1}\mathfrak{A} \subseteq \mathscr{S}_\alpha^{-1}\mathscr{Q}\mathfrak{A}$.

15. Suppose that $\mathbf{A}$ is a partial algebra of type $\tau \in \omega^\rho$ (so that all operations of $\mathbf{A}$ are finitary). Let $\{B_i \mid i \in I\} \subseteq \Sigma(\mathbf{A})$ be totally ordered by inclusion; that is, for every i and j, either $B_i \subseteq B_j$, or $B_j \subseteq B_i$. Prove

(a) $\bigcup_{i \in I} B_i \in \Sigma(\mathbf{A})$.
(b) If $\mathbf{B} = \mathbf{A} \restriction \bigcup_{i \in I} B_i$, and $\mathbf{B}_i = \mathbf{A} \restriction B_i$, then $\mathscr{S}_\omega\{\mathbf{B}\} = \bigcup_{i \in I} \mathscr{S}_\omega\{\mathbf{B}_i\}$.

16. Let $\mathfrak{A}$ be a class of partial algebras of type $\tau \in \omega^\rho$. Suppose that $\mathbf{A}$ is a partial algebra of type τ such that $\mathscr{S}_\omega\{\mathbf{A}\} \cap \mathscr{I}\mathfrak{A} \neq \varnothing$. Prove that there is a subalgebra $\mathbf{B}$ of $\mathbf{A}$ that is maximal with the property that $\mathbf{B} \in \mathscr{S}_\omega^{-1}\mathfrak{A}$. Show that $\mathbf{B}$ is unique (that is, a maximum subalgebra in $\mathscr{S}_\omega^{-1}\mathfrak{A}$) if the following amalgamation condition is satisfied:

$$\mathbf{C}, \mathbf{D} \in \mathfrak{A} \cap \mathscr{S}_\omega\{\mathbf{A}\} \quad \text{implies} \quad \mathbf{A} \restriction [C \cup D] \in \mathfrak{A}.$$

(*Hint:* show that if B_0 and B_1 are in $\Sigma(\mathbf{A})$, and if Y is a finite subset of $[B_0 \cup B_1]$, then there exists $X_0 \subseteq B_0$ and $X_1 \subseteq B_1$ with $|X_0| < \aleph_0$, $|X_1| < \aleph_0$, such that $Y \subseteq [X_0 \cup X_1]$.)

Notes to Chapter I

The basic definitions and results of this chapter are due to Birkhoff (see [1] and [3]). Cohn [9] presents the ideas in the more general context of categories of abstract algebras (with finitary operations). Other general references on abstract algebras are Grätzer [21], Kuroš [34], and Neumann [40].

[2]

Subdirect Decompositions

1. Subdirect Decompositions

The structure theory of algebras is concerned with the relation of arbitrary algebras to other algebras that enjoy special properties. The study of subdirect decompositions of an algebra is a special case of this problem. A subdirect product of a set $\{\mathbf{A}_i \mid i \in I\}$ of algebras is a certain kind of subalgebra of the direct product $\Pi_{i\in I}\mathbf{A}_i$. Our aim is to find a way to represent a given algebra as a subdirect product of a set $\{\mathbf{A}_i \mid i \in I\}$, where the algebras $\mathbf{A}_i$ have some property that makes them more tractable than arbitrarily given algebras. The usefulness of these representations will be shown in the final section of this chapter where the general results are applied to specific algebraic systems.

1.1 Definition. Let $\{\mathbf{A}_i \mid i \in I\}$ be a set (possibly empty) of relational systems of type $\tau \in \mathrm{Ord}^\rho$. Let $\Pi_{i\in I}\mathbf{A}_i = \langle B; S_\xi\rangle_{\xi<\rho}$. A *subdirect product* of $\{\mathbf{A}_i \mid i \in I\}$ is a relational system

$$\mathbf{A} = (\Pi_{i\in I}\mathbf{A}_i) \restriction X,$$

where $X \subseteq B$ is such that each of the mappings $\pi_i \restriction X$ is an epimorphism. The subdirect product is called *proper* if none of the mappings $\pi_i \restriction X$ is a monomorphism.

A *decomposition* of a relational system $\mathbf{A}$ of type τ as a subdirect product of $\{\mathbf{A}_i \mid i \in I\}$ is a monomorphism φ of $\mathbf{A}$ to $\Pi_{i\in I}\mathbf{A}_i$ such that $\pi_i \circ \varphi$ is an epimorphism for each $i \in I$. The decomposition is called *proper* if none of the mappings $\pi_i \circ \varphi$ is a monomorphism.

If $\varphi : \mathbf{A} \to \Pi_{i\in I}\mathbf{A}_i$ is a decomposition of $\mathbf{A}$ as a subdirect product of $\{\mathbf{A}_i \mid i \in I\}$, it is convenient to speak of *the subdirect decomposition* $\langle\varphi; \Pi_{i\in I}\mathbf{A}_i\rangle$ of $\mathbf{A}$. Clearly, if $\langle\varphi; \Pi_{i\in I}\mathbf{A}_i\rangle$ is a subdirect decomposition of $\mathbf{A}$, then $(\Pi_{i\in I}\mathbf{A}_i) \restriction \varphi(A)$ is a subdirect product of $\{\mathbf{A}_i \mid i \in I\}$, and the decomposition is proper if and only if this subdirect product is proper. However, in general $\mathbf{A}$ is not isomorphic to $(\Pi_{i\in I}\mathbf{A}_i) \restriction \varphi(A)$ (see Problem 2). In case $\mathbf{A}$ is an algebra, the monomorphism φ does map $\mathbf{A}$ isomorphically onto a subalgebra of the product by 1.4.10 and 1.2.4.

1.2 PROPOSITION. Let $\mathbf{A}$ be an algebra, and suppose that $\langle \varphi; \Pi_{i\in I}\mathbf{A}_i\rangle$ is a subdirect decomposition of $\mathbf{A}$. Let $\mathbf{B} = (\Pi_{i\in I}\mathbf{A}_i) \restriction \varphi(A)$. Then $\mathbf{B}$ is a subalgebra of $\Pi_{i\in I}\mathbf{A}_i$, and φ is an isomorphism of $\mathbf{A}$ to $\mathbf{B}$.

Because of this result, subdirect decompositions are of the most interest when applied to abstract algebras.

1.3 PROPOSITION. Let $\langle \varphi; \Pi_{i\in I}\mathbf{A}_i\rangle$ be a subdirect decomposition of a partial algebra $\mathbf{A}$. For each $i \in I$, let Γ_i be the kernel of the homomorphism $\pi_i \circ \varphi$. Then $\mathbf{A}_i$ is isomorphic to $\mathbf{A}/\Gamma_i$. Moreover, $\bigcap_{i\in I}\Gamma_i = I_A$. Finally, the decomposition is proper if and only if $\Gamma_i \supset I_A$ for all $i \in I$.

Proof. Since $\pi_i \circ \varphi$ is an epimorphism by hypothesis, it follows from 1.3.10 that $\mathbf{A}_i$ is isomorphic to $\mathbf{A}/\Gamma_i$. If $\langle x, y\rangle \in \bigcap_{i\in I}\Gamma_i$, then $\pi_i(\varphi(x)) = \pi_i(\varphi(y))$ for all $i \in I$. Therefore, $\varphi(x) = \varphi(y)$. Since φ is a monomorphism, it follows that $x = y$. Hence $\langle x, y\rangle \in I_A$. The last statement is obvious.

The converse of 1.3 is a much more important result for our purposes.

1.4 PROPOSITION. Let $\mathbf{A}$ be a partial algebra. Let $\{\Gamma_i \mid i \in I\}$ be a nonempty subset of $\Theta(\mathbf{A})$ such that $\bigcap_{i\in I}\Gamma_i = I_A$. Then there is a decomposition φ of $\mathbf{A}$ as a subdirect product of $\{\mathbf{A}/\Gamma_i \mid i \in I\}$ such that Γ_i is the kernel of $\pi_i \circ \varphi$.

Proof. By 1.5.6 there is a homomorphism φ of $\mathbf{A}$ to $\Pi_{i\in I}\mathbf{A}_i$ such that $\bar{\Gamma}_i = \pi_i \circ \varphi$ for all $i \in I$, and $\Gamma_\varphi = \bigcap_{i\in I}\Gamma_i = I_A$. Thus, φ is a monomorphism. Since $\bar{\Gamma}_i = \pi_i \circ \varphi$ is an epimorphism, it follows that $\langle \varphi; \Pi_{i\in I}\mathbf{A}/\Gamma_i\rangle$ is a decomposition of $\mathbf{A}$ as a subdirect product of $\{\mathbf{A}/\Gamma_i \mid i \in I\}$. Moreover Γ_i is the kernel of $\bar{\Gamma}_i = \pi_i \circ \varphi$.

1.5 DEFINITION. A partial algebra $\mathbf{A}$ is said to be subdirectly irreducible if $\mathbf{A}$ contains more than one element and there is no proper decomposition of $\mathbf{A}$ as a subdirect product.

1.6 LEMMA. A partial algebra $\mathbf{A}$ is subdirectly irreducible if and only if $\bigcap\{\Gamma \in \Theta(\mathbf{A}) \mid \Gamma \supset I_A\} \supset I_A$.

Proof. Let $\Gamma_0 = \bigcap\{\Gamma \in \Theta(\mathbf{A}) \mid \Gamma \supset I_A\}$. Assume that $\Gamma_0 \supset I_A$. Suppose that $\langle \varphi; \Pi_{i\in I}\mathbf{A}_i\rangle$ is a proper subdirect decomposition of $\mathbf{A}$. Let Γ_i be the kernel of $\pi_i \circ \varphi$. By 1.3, $\Gamma_i \supset I_A$ and $\bigcap_{i\in I}\Gamma_i = I_A$. Hence

$$I_A \subset \bigcap\{\Gamma \in \Theta(\mathbf{A}) \mid \Gamma \supset I_A\} \subseteq \bigcap_{i\in I}\Gamma_i = I_A,$$

which is a contradiction. Thus, $\mathbf{A}$ is subdirectly irreducible. Conversely, if $\Gamma_0 = I_A$, then $\mathbf{A}$ admits a proper subdirect decomposition by 1.4.

An equivalent statement of this lemma is that $\mathbf{A}$ is subdirectly irreducible if and only if there is a congruence relation $\Gamma_0 \supset I_A$ such that if $\Gamma \in \Theta(\mathbf{A})$ and $\Gamma \neq I_A$, then $\Gamma \supseteq \Gamma_0$. This version of 1.6 will be used in section 4 to characterize certain subdirectly irreducible algebras.

If **A** is an algebra that is not subdirectly irreducible, then it is possible to decompose **A** as a subdirect product of "smaller" algebras. If these algebras are not subdirectly irreducible, then they can be decomposed and so on. Ultimately, it might be hoped that **A** would be decomposed as a subdirect product of irreducible algebras. Mathematical experience leads us to believe that this hope is naive. However, we will show that in one important case the hope is not in vain—namely, when all of the operations of **A** are finitary.

A type τ is called *finitary* if $\tau \in \omega^J$ for some set J, that is, $\tau(j) < \omega$ for all $j \in J$. A relational system is said to be *finitary* if its type is finitary. To conclude this section, we will prove a basic property of the congruence lattices of finitary partial algebras.

1.7 LEMMA. Let **A** be a finitary partial algebra. Suppose that $\mathfrak{D} \subseteq \Theta(\mathbf{A})$ is a nonempty set that satisfies the following condition:

$$\varnothing \neq \{\Gamma_0, \Gamma_1, \cdots, \Gamma_{n-1}\} \subseteq \mathfrak{D}$$

implies that there exists $\Gamma \in \mathfrak{D}$ such that $\Gamma \supseteq \Gamma_i$ for all $i < n$. Then $\bigcup\mathfrak{D} \in \Theta(\mathbf{A})$.

Proof. Suppose that $\mathbf{A} = \langle A; F_\xi \rangle_{\xi<\rho}$. Define $\Delta = \bigcup\mathfrak{D}$. Since $\mathfrak{D} \neq \varnothing$, $\Delta \supseteq \Gamma$ for some $\Gamma \in \mathfrak{D}$. Thus, $I_A \subseteq \Gamma \subseteq \Delta$. Clearly, $\langle x, y \rangle \in \Delta$ implies $\langle y, x \rangle \in \Delta$. If $\langle x, y \rangle \in \Delta$ and $\langle y, z \rangle \in \Delta$, then $\langle x, y \rangle \in \Gamma_0 \in \mathfrak{D}$ and $\langle y, z \rangle \in \Gamma_1 \in \mathfrak{D}$. By the hypothesis on $\mathfrak{D}$, there exists $\Gamma \in \mathfrak{D}$ such that $\Gamma_0 \subseteq \Gamma$ and $\Gamma_1 \subseteq \Gamma$. Consequently, $\langle x, y \rangle \in \Gamma$, $\langle y, z \rangle \in \Gamma$, so that $\langle x, z \rangle \in \Gamma \subseteq \Delta$. This shows that Δ is an equivalence relation. Suppose that $\mathbf{a} \in \mathscr{D}(F_\xi)$, $\mathbf{b} \in \mathscr{D}(F_\xi)$, and $\langle \mathbf{a}(\eta), \mathbf{b}(\eta) \rangle \in \Delta$ for all $\eta < \tau(\xi)$. By definition of Δ, there exists $\Gamma_\eta \in \mathfrak{D}$ with $\langle \mathbf{a}(\eta), \mathbf{b}(\eta) \rangle \in \Gamma_\eta$ for each $\eta < \tau(\xi)$. Since $\tau(\xi) < \omega$, the hypothesis on $\mathfrak{D}$ implies that there exists $\Gamma \in \mathfrak{D}$ such that $\Gamma_\eta \subseteq \Gamma$ for all $\eta < \tau(\xi)$. Hence, $\langle \mathbf{a}(\eta), \mathbf{b}(\eta) \rangle \in \Gamma$ for all $\eta < \tau(\xi)$. Since Γ is a congruence relation, it follows that $\langle F_\xi(\mathbf{a}), F_\xi(\mathbf{b}) \rangle \in \Gamma \subseteq \Delta$. Therefore, Δ is a congruence relation.

2. Lattices

The study of subdirect decompositions (and even more so, direct decompositions) is greatly simplified when it is cast in the formalism of lattice theory. In this section, the basic concepts of lattice theory will be introduced.

We recall that a partially ordered set is a relational system $\mathbf{P} = \langle P; \leq \rangle$ with a binary relation that satisfies the conditions

(i) $a \leq a$ for all $a \in P$.
(ii) $a \leq b$ and $b \leq a$ implies $a = b$.
(iii) $a \leq b$ and $b \leq c$ implies $a \leq c$.

The most important example of a partially ordered set occurs when $P \subseteq \mathbf{P}(X)$ for some set X, and $\leq$ is set inclusion. Note that $\Sigma(\mathbf{A})$ and $\Theta(\mathbf{A})$ are both examples of this kind: if $\mathbf{A} = \langle A; F_\xi \rangle_{\xi < \rho}$, then $\Sigma(\mathbf{A}) \subseteq \mathbf{P}(A)$ and $\Theta(\mathbf{A}) \subseteq \mathbf{P}(A \times A)$. Hence $\Sigma(\mathbf{A})$ and $\Theta(\mathbf{A})$ enjoy a natural partial ordering, and we will always consider them as partially ordered sets.

In talking about partially ordered sets, it is frequently convenient to turn the inequality symbol around, writing $b \geq a$ instead of $a \leq b$. It is not absolutely necessary to allow this convention, but sometimes it is desirable in order to shift the emphasis in a statement or formula. Another useful notational device is that of writing $a \nleq b$ or $b \ngeq a$ for the negation of the statement $a \leq b$. Finally, the relation of *strict inequality* is defined by

$$a < b \text{ (or } b > a) \quad \text{if} \quad a \leq b \quad \text{and} \quad a \neq b.$$

2.1 DEFINITION. Let $\mathbf{P}$ be a partially ordered set. Let $A \subseteq P$. An element $b \in P$ is an *upper bound* of A if $a \leq b$ for all $a \in A$. An element $c \in P$ is called a *lower bound* of A if $c \leq a$ for all $a \in A$. An element $d \in P$ is called a *least upper bound* of A in $\mathbf{P}$ if

(i) d is an upper bound of A, and
(ii) if b is an upper bound of A, then $d \leq b$.

An element $e \in P$ is called a *greatest lower bound* of A in $\mathbf{P}$ if

(i′) e is a lower bound of A, and
(ii′) if c is a lower bound of A, then $c \leq e$.

It is conceivable that with a suitable choice of partially ordered set $\mathbf{P}$ and set $A \subseteq P$, any one of the following situations might prevail:

(1) A has no least upper bound in $\mathbf{P}$.
(2) A has exactly one least upper bound in $\mathbf{P}$.
(3) A has more than one least upper bound in $\mathbf{P}$.

We will show that the third alternative is impossible.

2.2 LEMMA. Let A be a subset of a partially ordered set $\mathbf{P}$. Then A has at most one least upper bound in $\mathbf{P}$ and at most one greatest lower bound in $\mathbf{P}$.

Proof. Let d and e be least upper bounds of A. Condition (i) of 2.1 implies that d is an upper bound of A. Therefore, since e satisfies condition (ii) of 2.1, it follows that $e \leq d$. By symmetry, $d \leq e$. Hence, $d = e$. A similar argument shows that there is at most one greatest lower bound.

Either one of the conditions (1) and (2) listed above can occur. A given set A need not have any least upper bound. In fact, it is easy to give examples of sets which have no upper bound at all.

A particular case of definition 2.1 is worth noting. If A is the empty set, then every element of P is both an upper bound of A and a lower bound of A. Hence, a least upper bound of $\varnothing$ is an element o satisfying $o \leq a$ for all

$a \in P$. That is, $\varnothing$ has a least upper bound in **P** if and only if **P** has a smallest element o. Similarly, $\varnothing$ has a greatest lower bound if and only if **P** has a largest element u.

The most common examples of least upper bounds and greatest lower bounds occur for sets of sets as described in the next result.

2.3 LEMMA. Let X be any set, and suppose that $\mathscr{L} \subseteq \mathbf{P}(X)$. Suppose that $\mathfrak{A} \subseteq \mathscr{L}$ is such that $\bigcup\mathfrak{A} \in \mathscr{L}$. Then $\bigcup\mathfrak{A}$ is the least upper bound of $\mathfrak{A}$ in the partially ordered set $\langle\mathscr{L}; \subseteq\rangle$. If $\mathfrak{B} \subseteq \mathscr{L}$ is such that $\bigcap\mathfrak{B} \in \mathscr{L}$, then $\bigcap\mathfrak{B}$ is the greatest lower bound of $\mathfrak{B}$.

Proof. If $A \in \mathfrak{A}$, then $A \subseteq \bigcup\mathfrak{A}$. Hence, $\bigcup\mathfrak{A}$ is an upper bound of $\mathfrak{A}$. If $B \in \mathscr{L}$ is any upper bound of $\mathfrak{A}$, then $A \subseteq B$ for all $A \in \mathfrak{A}$. Therefore, $\bigcup\mathfrak{A} \subseteq B$. This shows that $\bigcup\mathfrak{A}$ satisfies the definition of a least upper bound of $\mathfrak{A}$. The proof of the second statement is similar.

It should be emphasized that a subset $\mathfrak{A}$ of a partially ordered set $\mathscr{L} \subseteq \mathbf{P}(X)$ may have a least upper bound in $\mathscr{L}$ which is different from $\bigcup\mathfrak{A}$. Similarly, the greatest lower bound other than $\bigcap\mathfrak{A}$ may exist in $\mathscr{L}$. Of course, in these respective cases $\bigcup\mathfrak{A}$ and $\bigcap\mathfrak{A}$ do not belong to $\mathscr{L}$. A *complete lattice* is a partially ordered set in which every subset has a least upper bound and a greatest lower bound.

It is convenient to introduce notation for the least upper bound and greatest lower bound of a subset of a partially ordered set. If A is a subset of a partially ordered set **P**, we will write $\bigvee A$ for the least upper bound of A, provided it exists, and $\bigwedge A$ for the greatest lower bound of A when it exists. Thus, $\bigvee A$ not only designates an element of P, but the act of writing this expression tacitly implies that A has a least upper bound in **P**. A similar remark applies to $\bigwedge A$. Of course, if **P** is a complete lattice, then $\bigvee A$ and $\bigwedge A$ are defined for all $A \subseteq P$.

In the case of two element sets, it is convenient to use operational notation:

$$a \vee b = \bigvee\{a, b\},$$

$$a \wedge b = \bigwedge\{a, b\}.$$

Thinking of $\vee$ and $\wedge$ in this way as binary operations, a partially ordered set is a partial algebra of type $\tau = \{\langle 0, 2\rangle, \langle 1, 2\rangle\}$. A *lattice* is a partially ordered set for which the partial operations $\vee$ and $\wedge$ are actually operations, that is, each pair of elements has a least upper and a greatest lower bound.

The partial operations $\vee$ and $\wedge$ are usually called the "join" and "meet," respectively. It is customary to use expressions such as "the join of a and b" or "a join b" to designate $a \vee b$ verbally. In a similar way, the expressions "the meet of a and b" or "a meet b" are used for $a \wedge b$.

It is important that lattices can be considered as abstract algebras. They are in fact characterized by a convenient set of identities.

2.5 Proposition. Let **L** be a lattice. Then the following are identically satisfied for all a, b, and c in L:

(i) $a \vee a = a$ and $a \wedge a = a$.
(ii) $a \vee b = b \vee a$ and $a \wedge b = b \wedge a$.
(iii) $a \vee (b \vee c) = (a \vee b) \vee c$ and $a \wedge (b \wedge c) = (a \wedge b) \wedge c$.
(iv) $a \wedge (b \vee a) = a$ and $a \vee (b \wedge a) = a$.

The partial ordering of **L** is determined by the join or meet operation of **L**. In fact, for any a and b in L, the following conditions are equivalent:

$$a \leq b, \qquad a \wedge b = a, \qquad a \vee b = b.$$

Conversely, if **L** is an algebra with two binary operations $\vee$ and $\wedge$ satisfying (i)–(iv), then the conditions $a \wedge b = a$ and $a \vee b = b$ are equivalent, and if the binary relation $\leq$ is defined to be the set $\{\langle a, b\rangle \mid a \wedge b = a\}$, then $\langle L; \leq\rangle$ is a lattice in which $a \vee b$ is the least upper bound of $\{a, b\}$ and $a \wedge b$ is the greatest lower bound of $\{a, b\}$.

Proof. It is obvious that (i) and (ii) are satisfied by the definition of the least upper bound and greatest lower bound. To prove (iii), let $d = a \vee (b \vee c)$. Then $a \leq d$ and $b \vee c \leq d$. Consequently, $b \leq d$ and $c \leq d$. Thus, d is an upper bound of $\{a, b, c\}$. Moreover, if e is an upper bound of $\{a, b, c\}$, then $e \geq a$ and e is an upper bound of $\{b, c\}$. Thus, $b \vee c \leq e$. Consequently, $d = a \vee (b \vee c) \leq e$. This proves that $d = a \vee (b \vee c)$ is the least upper bound of $\{a, b, c\}$. Similarly, $(a \vee b) \vee c$ is the least upper bound of $\{a, b, c\}$. Therefore, $a \vee (b \vee c) = (a \vee b) \vee c$. By an analogous argument, $a \wedge (b \wedge c) = (a \wedge b) \wedge c$. To prove the identity $a \wedge (b \vee a) = a$, notice that $a \leq b \vee a$. Thus, a is a lower bound of $\{a, b \vee a\}$. Moreover, any lower bound of $\{a, b \vee a\}$ is $\leq a$. Therefore, $a = a \wedge (b \vee a)$. By a similar proof, $a = a \vee (b \wedge a)$. We next show that the identities of (ii) and (iv) yield the equivalence of $a \wedge b = a$ and $a \vee b = b$. Indeed, if $a \wedge b = a$, then by (ii) and (iv), $a \vee b = (a \wedge b) \vee b = b \vee (a \wedge b) = b$. If $a \vee b = b$, then $a \wedge b = a \wedge (a \vee b) = a \wedge (b \vee a) = a$. Therefore, it follows from (ii) and (iv) that $a \wedge b = a$ and $a \vee b = b$ are equivalent conditions. If **L** is a lattice and $a \wedge b = a$, then $a \leq b$, because $a \wedge b \leq b$. If $a \leq b$, then a is a lower bound of $\{a, b\}$ and every lower bound of this set is $\leq a$. Thus, $a = a \wedge b$. This shows that the conditions $a \leq b$, $a \wedge b = a$, and $a \vee b = b$ are equivalent in a lattice.

Suppose now that **L** is an algebra with two binary operations satisfying (i)–(iv). It was shown above that the conditions $a \wedge b = a$ and $a \vee b = b$ are equivalent in **L**. Define $a \leq b$ if $a \wedge b = a$. By (i), $a \leq a$. If $a \leq b$ and $b \leq a$, then $a = a \wedge b = b \wedge a = b$ by (ii). Assume that $a \leq b$ and $b \leq c$. Then $a \wedge b = a$ and $b \wedge c = b$. Hence, by (iii), $a = a \wedge b = a \wedge (b \wedge c) = (a \wedge b) \wedge c = a \wedge c$. Therefore, $a \leq c$. Consequently, $\langle L; \leq\rangle$ is a partially ordered set. To complete the proof, it suffices to show that if $\{a, b\} \subseteq L$, then $a \vee b$ is the least upper bound of $\{a, b\}$, and $a \wedge b$ is the greatest lower bound of $\{a, b\}$. Since $b \wedge (a \vee b) = b$, it follows that $b \leq a \vee b$. From $a \wedge (a \vee b) = (a \vee b) \wedge a = a$, we obtain $a \leq a \vee b$. Hence, $a \vee b$ is an upper bound of $\{a, b\}$. If c is an upper bound of $\{a, b\}$, then $a \vee c = c$ and $b \vee c = c$. Therefore, $c = a \vee c = a \vee (b \vee c) = (a \vee b) \vee c$. Thus, $a \vee b \leq c$. This proves that $a \vee b$ is the least upper bound of $\{a, b\}$. By a similar argument, it follows that $a \wedge b$ is the greatest lower bound of $\{a, b\}$. This completes the proof of 2.5.

The proof given above that $a \vee (b \vee c) = \text{l.u.b.}\ \{a, b, c\}$ can be extended by induction to show that

$$a_0 \vee (a_1 \vee \cdots \vee (a_{n-1} \vee a_n) \cdots) = \text{l.u.b.}\ \{a_0, a_1, \cdots, a_{n-1}, a_n\}.$$

In particular, every finite, nonempty subset of a lattice has a least upper bound. Similarly, every finite, nonempty subset has a greatest lower bound.

If $\mathbf{L}_1$ and $\mathbf{L}_2$ are lattices which are isomorphic as partially ordered sets, then it is easy to see that they are isomorphic as abstract algebras. Conversely, if $\mathbf{L}_1$ and $\mathbf{L}_2$ are isomorphic as abstract algebras, then they are isomorphic as partially ordered sets. Moreover, a homomorphism of $\mathbf{L}_1$ to $\mathbf{L}_2$ considered as abstract algebras with the join and meet operations will preserve the ordering. However, a homomorphism of $\mathbf{L}_1$ to $\mathbf{L}_2$ considered as relational systems will generally not preserve the join and meet operations. Unless otherwise stated, a homomorphism from one lattice to another should be interpreted as an algebra homomorphism, that is, a homomorphism which commutes with the join and meet operations.

Most of the lattices that occur in the study of the structure of partial algebras and algebras are complete. In fact, the important lattices $\Sigma(\mathbf{A})$ and $\Theta(\mathbf{A})$ are of the type described in the next result.

2.6 PROPOSITION. Let X be any set. Suppose that $\mathscr{L} \subseteq \mathbf{P}(X)$ satisfies the condition

$$\mathfrak{A} \subseteq \mathscr{L} \quad \text{implies} \quad \bigcap \mathfrak{A} \in \mathscr{L}.$$

Then $\langle \mathscr{L}; \subseteq \rangle$ is a complete lattice.

Proof. Let $\mathfrak{A} \subseteq \mathscr{L}$. By assumption $\bigcap \mathfrak{A} \in \mathscr{L}$. Hence by 2.3 $\mathfrak{A}$ has a greatest lower bound in $\mathscr{L}$. To show that $\mathfrak{A}$ has a least upper bound, let $\mathfrak{A}^* = \{C \in \mathscr{L} \mid C \supseteq \bigcup \mathfrak{A}\}$. Define $B = \bigcap \mathfrak{A}^*$. By assumption, $B \in \mathscr{L}$. Clearly, $B = \bigcap \mathfrak{A}^* \supseteq \bigcup \mathfrak{A} \supseteq A$ for all $A \in \mathfrak{A}$. Hence, B is an upper bound of $\mathfrak{A}$. If C is an upper bound of $\mathfrak{A}$ in $\mathscr{L}$, then $C \supseteq \bigcup \mathfrak{A}$. Therefore $C \in \mathfrak{A}^*$, and it follows that $C \supseteq \bigcap \mathfrak{A}^* = B$. Consequently, B is the least upper bound of $\mathfrak{A}$.

It should be noted that the hypothesis in 2.6 implies that $X \in \mathscr{L}$. In fact, X is the intersection of the empty subset of $\mathscr{L}$.

Together with 1.3.5 and 1.4.4, Proposition 2.6 implies that the partially ordered sets $\Sigma(\mathbf{A})$ and $\Theta(\mathbf{A})$ are complete lattices for any partial algebra $\mathbf{A}$. Unfortunately, it is not possible to obtain much information about the structure theory of an algebra $\mathbf{A}$ from the mere fact that $\Sigma(\mathbf{A})$ and $\Theta(\mathbf{A})$ are complete lattices. However, the lattices of subalgebras and congruence relations of a finitary algebra satisfy a stronger condition that has interesting consequences.

2.7 DEFINITION. A subset A of a partially ordered set $\mathbf{P}$ is said to be directed if every finite subset of A has an upper bound in A.

In particular, a directed subset A of $\mathbf{P}$ cannot be empty, since the empty set is a finite subset of A which must have an upper bound in A. It is easy to prove by induction that A is directed if and only if $A \neq \varnothing$ and every pair of elements in A has an upper bound in A.

2.8 DEFINITION. Let $\mathbf{L}$ be a complete lattice. An element $c \in L$ is called *compact* if, whenever $D \subseteq L$ is a directed set which satisfies $c \leq \bigvee D$, then $c \leq a$ for some $a \in D$. A lattice $\mathbf{L}$ is *compactly generated* if it is complete and every element of L is the least upper bound of a set of compact elements of L.

This is a rather complicated definition, and it is surprising that compactly generated lattices occur naturally.

2.9 PROPOSITION. Let X be any set. Suppose that $\mathscr{L} \subseteq \mathbf{P}(X)$ satisfies the conditions

(i) $\mathfrak{A} \subseteq \mathscr{L}$ implies $\bigcap \mathfrak{A} \in \mathscr{L}$.
(ii) $\mathfrak{D} \subseteq \mathscr{L}$, $\mathfrak{D}$ directed implies $\bigcup \mathfrak{D} \in \mathscr{L}$.

Then $\langle \mathscr{L}; \subseteq \rangle$ is a compactly generated lattice.

Proof. By 2.6, $\mathscr{L}$ is a complete lattice. Moreover, if $\mathfrak{D}$ is a directed subset of $\mathscr{L}$, then $\bigvee \mathfrak{D} = \bigcup \mathfrak{D}$ by 2.3. Let $x \in X$. Define $[x] = \bigcap \{A \in \mathscr{L} \mid x \in A\}$. It will be shown that $[x]$ is compact. Suppose that $[x] \subseteq \bigvee \mathfrak{D}$, where $\mathfrak{D} \subseteq \mathscr{L}$ and $\mathfrak{D}$ is directed. Then $x \in [x] \subseteq \bigcup \mathfrak{D}$. Therefore, $x \in D$ for some $D \in \mathfrak{D}$. Consequently, $[x] \subseteq D$, by the definition of $[x]$. This shows that $[x]$ is compact. Finally, if A is any set of $\mathscr{L}$, then

$$A = \bigcup \{\{x\} \mid x \in A\} \subseteq \bigcup \{[x] \mid x \in A\} \subseteq \bigvee \{[x] \mid x \in A\} \subseteq A.$$

The last inclusion follows from the fact that if $x \in A$, then $[x] \subseteq A$, so that A is an upper bound of $\{[x] \mid x \in A\}$. This proves that A is a join of compact elements. Therefore, $\mathscr{L}$ is compactly generated.

2.10 COROLLARY. Let $\mathbf{A}$ be a finitary partial algebra. Then $\Theta(\mathbf{A})$ is a compactly generated lattice.

This corollary follows immediately from 2.9, 1.3.5, and 1.7.

3. The Subdirect Decomposition Theorem

The principal theorem of this chapter will now be presented.

3.1 DEFINITION. Let **L** be a complete lattice. An element $m \in L$ is called *meet irreducible** if

$$m < \bigwedge\{x \in L \mid x > m\}.$$

Equivalently, m is meet irreducible if $m = \bigwedge\{x_i \mid i \in I\}$ implies $x_i = m$ for some i.

We have no reason to expect that an arbitrary complete lattice has any meet irreducible elements other than its greatest element. For example, the set U of real numbers in the unit interval $[0, 1]$, given the usual ordering, is a complete lattice for which 1 is the only element which is meet irreducible. However, in compactly generated lattices the meet irreducible elements are abundant. The following theorem proves this assertion.

3.2 THEOREM. Let **L** be a compactly generated lattice. Then every element of **L** is the meet of the set of all meet irreducible elements that contain it. That is, if M is the set of all meet irreducible elements of **L**, then for every $a \in L$,

$$a = \bigwedge\{m \in M \mid m \geq a\}.$$

Proof. Let $b = \bigwedge\{m \in M \mid m \geq a\}$. Clearly, $a \leq b$. Suppose that $a < b$. Since **L** is compactly generated, b is the least upper bound of a set of compact elements. Consequently, there is a compact element c such that $c \leq b$ and $c \nleq a$. We will prove:

(i) There exists $m \in M$ such that $a \leq m$ and $c \nleq m$.

This will yield a contradiction, because by the definition of b, if $m \in M$ and $m \geq a$, then $m \geq b \geq c$. To prove (i), let $\mathfrak{D}$ be the set of all $D \subseteq L$ that satisfies

(ii) D is a directed subset of L.

(iii) $d \in D$ implies $a \leq d$, $c \nleq d$.

Clearly, $\{a\} \in \mathfrak{D}$. Moreover, if $\mathfrak{C}$ is a simply ordered subset of $\mathfrak{D}$, then it is easy to see that $\bigcup\mathfrak{C}$ satisfies (ii) and (iii). Consequently, $\mathfrak{D}$ satisfies the conditions of Zorn's lemma, so that there is a maximal set $D \in \mathfrak{D}$. Let $m = \bigvee D$. Since $D \neq \varnothing$, it follows from (iii) that $m \geq a$. If $c \leq m = \bigvee D$, then $c \leq d$ for some $d \in D$ by the compactness of c. This contradicts

* The usual terminology of lattice theory for this concept is *completely meet irreducible.* However, since the standard notion of meet irreducibility will not be used, it is convenient to adopt the shorter term "meet irreducible."

condition (iii). Therefore $c \nleq m$. The proof is completed by showing that m is meet irreducible. Since $c \nleq m$, it follows that m is not the largest element of **L**. Suppose that $m < e$. Then $e \notin D$, so that $D \cup \{e\} \supset D$. By the maximality of D, we have $D \cup \{e\} \notin \mathfrak{D}$. However, $D \cup \{e\}$ is a non-empty directed set in L, and every member of $D \cup \{e\}$ is $\geq a$. Moreover, if $d \in D$, then $c \nleq d$. Therefore, the only explanation for $D \cup \{e\}$ not being in $\mathfrak{D}$ is that $c \leq e$. This shows that if $m < e$, then $e \geq m \vee c$. Thus, $\bigwedge\{e \in L \mid e > m\} \geq m \vee c > m$. Therefore, m is meet irreducible, and the proof is complete.

A lemma is needed in the proof of the main theorem on the existence of subdirect decompositions. This result is of general interest.

3.3 Proposition. Let **A** and **B** be similar partial algebras. Suppose that $\varphi \in \text{Hom}(\mathbf{A}, \mathbf{B})$ is an epimorphism. Then $\Theta(\mathbf{B})$ is isomorphic (as a lattice) to $\{\Gamma \in \Theta(\mathbf{A}) \mid \Gamma \supseteq \Gamma_\varphi\}$.

Proof. For $\Delta \in \Theta(\mathbf{B})$, let $\gamma(\Delta)$ denote the kernel of the epimorphism $\bar{\Delta} \circ \varphi$. It is clear that $\gamma(\Delta) \supseteq \Gamma_\varphi$, and if $\Delta_1 \subseteq \Delta_2$, then $\gamma(\Delta_1) \subseteq \gamma(\Delta_2)$. On the other hand, suppose that $\gamma(\Delta_1) \subseteq \gamma(\Delta_2)$. Let $\langle c, d\rangle \in \Delta_1$. Since φ is an epimorphism, there are elements a and b such that $c = \varphi(a)$ and $d = \varphi(b)$. Hence $\bar{\Delta}_1(\varphi(a)) = \bar{\Delta}_1(c) = \bar{\Delta}_1(d) = \bar{\Delta}_1(\varphi(b))$. Thus, $\langle a, b\rangle \in \gamma(\Delta_1) \subseteq \gamma(\Delta_2)$, so that $\bar{\Delta}_2(c) = \bar{\Delta}_2(\varphi(a)) = \bar{\Delta}_2(\varphi(b)) = \bar{\Delta}_2(d)$. Consequently, $\langle c, d\rangle \in \Delta_2$. This shows that $\Delta_1 \subseteq \Delta_2$. It remains only to prove that if $\Gamma \in \Theta(\mathbf{A})$ and $\Gamma \supseteq \Gamma_\varphi$, then $\Gamma = \gamma(\Delta)$ for some $\Delta \in \Theta(\mathbf{B})$. By 1.3.7, $\Gamma \supseteq \Gamma_\varphi$ implies that there is a homomorphism ψ of **B** to $\mathbf{A}/\Gamma$ such that $\bar{\Gamma} = \psi \circ \varphi$. Let Δ be the kernel of ψ. Since $\langle a, b\rangle \in \Gamma$ is equivalent to $\psi(\varphi(a)) = \bar{\Gamma}(a) = \bar{\Gamma}(b) = \psi(\varphi(b))$, it follows that $\langle a, b\rangle \in \Gamma$ if and only if $\langle \varphi(a), \varphi(b)\rangle \in \Delta$. Hence, Γ is the kernel of $\bar{\Delta} \circ \varphi$, that is $\gamma(\Delta)$. The proof is complete.

3.4 Theorem. Let **A** be a finitary partial algebra. Then **A** admits a decomposition as a subdirect product of partial algebras that are subdirectly irreducible.

Proof. By 2.10, $\Theta(\mathbf{A})$ is compactly generated. Therefore, by 3.2 it is possible to find a set $\{\Gamma_i \mid i \in I\} \subseteq \Theta(\mathbf{A})$ such that each Γ_i is meet irreducible and $\bigcap_{i\in I}\Gamma_i = I_A$. By 1.4, **A** admits a decomposition as a subdirect product of $\{\mathbf{A}/\Gamma_i \mid i \in I\}$. By 3.3, $\Theta(\mathbf{A}/\Gamma_i)$ is isomorphic to $\{\Gamma \in \Theta(\mathbf{A}) \mid \Gamma \supseteq \Gamma_i\}$. Since Γ_i is meet irreducible in $\Theta(\mathbf{A})$, it follows that I_{A/Γ_i} is meet irreducible in $\Theta(\mathbf{A}/\Gamma_i)$. Hence, $\mathbf{A}/\Gamma_i$ is subdirectly irreducible by 1.6.

The most important case of 3.4 is when **A** is an abstract algebra. In this situation, the statement of 3.4 can be strengthened, using 1.2.4 and 1.4.10.

3.5 Corollary. Let **A** be a finitary abstract algebra. Then **A** is isomorphic to a subdirect product

$$\mathbf{B} = (\Pi_{i\in I}\mathbf{A}_i) \restriction X,$$

where each $\mathbf{A}_i$ is subdirectly irreducible and **B** is a subalgebra of $\Pi_{i\in I}\mathbf{A}_i$.

4. Applications

In this section, Theorem 3.4 will be used to obtain representation theorems for two particular types of abstract algebras—Boolean algebras and Abelian groups. These are common classes of algebras, and each is defined by a certain set of identities.† It is easily seen that under these circumstances the classes are closed under the operations of taking epimorphisms, subalgebras, and products. In such cases, Theorem 3.4 provides a characterization of the class in terms of the algebras (in the class) that are subdirectly irreducible.

4.1 PROPOSITION. Let $\mathfrak{A}$ be a class of abstract algebras of finitary type τ such that $\mathscr{Q}\mathfrak{A} = \mathscr{S}\mathfrak{A} = \mathscr{P}\mathfrak{A} = \mathfrak{A}$. Then an algebra **A** of type τ is in $\mathfrak{A}$ if and only if **A** is isomorphic to a subalgebra of a direct product of subdirectly irreducible algebras in $\mathfrak{A}$.

This proposition shows that for classes defined by a set of identities, it is decidedly worthwhile to have a characterization of the subdirectly irreducible algebras. This is what will be done in this section for the classes of Boolean algebras and Abelian groups.

4.2 EXAMPLE. A *Boolean algebra* $\mathbf{B} = \langle B; \vee, \wedge, o, u, ' \rangle$ is an algebra of type $\{\langle 0, 2\rangle, \langle 1, 2\rangle, \langle 2, 0\rangle, \langle 3, 0\rangle, \langle 4, 1\rangle\}$ such that $\langle B; \vee, \wedge, 0, u\rangle$ is a *distributive lattice* with least element o and greatest element u (that is, the binary operations $\vee$ and $\wedge$ satisfy the conditions (i), (ii), (iii), and (iv) of 2.5, together with the distributive laws $x \wedge (y \vee z) = (x \wedge y) \vee (x \wedge z)$ and $x \vee (y \wedge z) = (x \vee y) \wedge (x \vee z)$, and $x \wedge o = o, x \vee u = u, x \vee o = x, x \wedge u = x$ for all $x \in B$), and the unary operation $'$ satisfies

$$x \wedge x' = o, \qquad x \vee x' = u$$

for all $x \in B$. The element x' is called the *complement of* x *in* B, and the operation $'$ is called complementation.

The most common example of a Boolean algebra is $\langle \mathbf{P}(X); \cup, \cap, \varnothing, X, {}^c\rangle$, where X is any set, $\cup$, $\cap$ are the union and intersection operations, and ^{c} is the operation of complementation relative to X, that is, $Y^c = X - Y$ for $Y \subseteq X$. We will use the subdirect decomposition theorem to show that every Boolean algebra is isomorphic to a subalgebra of such a Boolean algebra of sets. In other words, it will be proved that every Boolean algebra is isomorphic to a Boolean algebra of the form $\langle \mathfrak{S}; \cup, \cap, \varnothing, X, {}^c\rangle$, where $\mathfrak{S} \subseteq \mathbf{P}(X)$, $\varnothing \in \mathfrak{S}$, $X \in \mathfrak{S}$, and for any A, B in $\mathfrak{S}$, $A \cup B \in \mathfrak{S}$, $A \cap B \in \mathfrak{S}$, and $A^c \in \mathfrak{S}$.

† The meaning of the term identity for algebras of a given type will be discussed more fully in Chapter 5.

Let **B** be any Boolean algebra, and suppose that $a \in B$. Define

$$\Gamma_a = \{\langle x, y\rangle \mid x \vee a = y \vee a\}.$$

It is obvious that Γ_a is an equivalence relation on **B**. Suppose that $\langle x, y\rangle \in \Gamma_a$ and $\langle z, w\rangle \in \Gamma_a$, that is, $x \vee a = y \vee a$ and $z \vee a = w \vee a$. Then $(x \vee z) \vee a = x \vee (z \vee a) = x \vee (w \vee a) = w \vee (x \vee a) = w \vee (y \vee a) = (y \vee w) \vee a$. Hence, $\langle x \vee z, y \vee w\rangle \in \Gamma_a$. Also, $(x \wedge z) \vee a = (x \vee a) \wedge (z \vee a) = (y \vee a) \wedge (w \vee a) = (y \wedge w) \vee a$, so that $\langle x \wedge z, y \wedge w\rangle \in \Gamma_a$. Suppose that $\langle x, y\rangle \in \Gamma_a$, that is, $x \vee a = y \vee a$. Then $x' \vee a = (x' \vee a) \vee (y \wedge y') = ((x' \vee a) \vee y) \wedge ((x' \vee a) \vee y') = (x' \vee (y \vee a)) \wedge (a \vee x' \vee y') = (x' \vee (x \vee a)) \wedge (a \vee x' \vee y') = (a \vee u) \wedge (a \vee x' \vee y') = u \wedge (a \vee x' \vee y') = a \vee x' \vee y'$. By symmetry, $y' \vee a = a \vee x' \vee y'$ also. Hence $\langle x', y'\rangle \in \Gamma_a$. We have therefore proved that Γ_a is a congruence relation on **B**.

Suppose that **B** is a subdirectly irreducible Boolean algebra. By 1.6 there is a congruence relation $\Gamma_0 \in \Theta(\mathbf{B})$ with $\Gamma_0 \supset I_B$, such that if $\Gamma \in \Theta(\mathbf{B})$ and $\Gamma \neq I_B$, then $\Gamma \supseteq \Gamma_0$. Let $a \in B$. We will prove that either $a = o$ or $a = u$. Suppose that $a \neq o$ and $a \neq u$. Then $a' \neq o$. Since $\langle o, a\rangle \in \Gamma_a$ and $\langle o, a'\rangle \in \Gamma_{a'}$, it follows that $\Gamma_a \neq I_B$ and $\Gamma_{a'} \neq I_B$. Hence, $\Gamma_a \supseteq \Gamma_0$ and $\Gamma_{a'} \supseteq \Gamma_0$. Let $\langle x, y\rangle \in \Gamma_0$, $x \neq y$. Then $\langle x, y\rangle \in \Gamma_a$ and $\langle x, y\rangle \in \Gamma_{a'}$. That is, $x \vee a = y \vee a$ and $x \vee a' = y \vee a'$. Consequently, $x = x \vee o = x \vee (a \wedge a') = (x \vee a) \wedge (x \vee a') = (y \vee a) \wedge (y \vee a') = y \vee (a \wedge a') = y \vee o = y$. This contradiction proves the assertion that $a = o$ or $a = u$. This argument shows that the subdirectly irreducible Boolean algebras are uniquely determined: $\mathbf{B} = \langle\{o, u\}; \vee, \wedge, o, u, '\rangle$, where $o \vee o = o$, $o \vee u = u \vee o = u \vee u = u$; $o \wedge o = o \wedge u = u \wedge o = o$, $u \wedge u = u$; $o' = u$, $u' = o$. It is customary to denote this two-element Boolean algebra by the symbol **2**.

Since Boolean algebras are defined to be abstract algebras of type $\{\langle 0, 2\rangle, \langle 1, 2\rangle, \langle 2, 0\rangle, \langle 3, 0\rangle, \langle 4, 1\rangle\}$, which satisfy certain identities, it follows that every epimorphic image of a Boolean algebra is a Boolean algebra. Hence, by 4.1 every Boolean algebra is isomorphic to a subalgebra of a direct product of two-element Boolean algebras. Suppose that $\Pi_{i\in I}\mathbf{B}_i$ is such a product, that is, each $\mathbf{B}_i$ contains only two elements o and u. Consider the mapping

$$\varphi : a \rightarrow \{i \in I \mid a(i) = u\}$$

of $\mathbf{X}_{i\in I}B_i$ to $\mathbf{P}(I)$. We will show that φ is an isomorphism of $\Pi_{i\in I}B_i$ to $\langle \mathbf{P}(I); \cup, \cap, \varnothing, I, {}^c\rangle$. Since $a \vee b$ is defined by $(a \vee b)(i) = a(i) \vee b(i)$, and in a two-element Boolean algebra $a(i) \vee b(i) = u$ if and only if either $a(i) = u$ or $b(i) = u$, it follows that $\varphi(a \vee b) = \{i \in I \mid a(i) = u\} \cup \{i \in I \mid b(i) = u\} = \varphi(a) \cup \varphi(b)$. Similarly, $\varphi(a \wedge b) = \varphi(a) \cap \varphi(b)$, $\varphi(o) = \varnothing$, $\varphi(u) = I$, and $\varphi(a') = \varphi(a)^c$. Thus, φ is a homomorphism. If $\varphi(a) = \varphi(b)$, then for all i, $a(i) = u$ if and only if $b(i) = u$. Thus, $a(i) = o$ if and only if $b(i) = o$. Therefore, $a = b$. That is, φ is a monomorphism. Finally, if $K \in \mathbf{P}(I)$, define $a \in \mathbf{X}_{i\in I}B_i$ by $a(i) = u$ if $i \in K$ and $a(i) = o$ if $i \notin K$. Clearly, $\varphi(a) = K$. Therefore, φ is an epimorphism. This completes the proof that $\Pi_{i\in I}\mathbf{B}_i$ is isomorphic to $\langle \mathbf{P}(I); \cup, \cap, \varnothing, I, {}^c\rangle$. It is now possible to interpret the result that every Boolean algebra admits a decomposition as a subdirect product of copies of **2** in the following way.

4.3 THEOREM. (Stone [48]) Every Boolean algebra is isomorphic to an algebra of sets, that is, an algebra of the form

$$\langle \mathfrak{S}; \cup, \cap, \varnothing, X, {}^c\rangle,$$

where $\mathfrak{S} \subseteq \mathbf{P}(X)$, and $\cup$, $\cap$, and ^{c} denote set theoretical union, intersection and complement respectively.

4.4 EXAMPLE. Let $\mathbf{A} = \langle A; +, 0, -\rangle$ be an Abelian group. That is, $\mathbf{A}$ is an algebra of type $\{\langle 0, 2\rangle, \langle 1, 0\rangle, \langle 2, 1\rangle\}$ satisfying the identities: $a + b = b + a, a + (b + c) = (a + b) + c, a + 0 = a$, and $a + (-a) = 0$. Our objective is to identify the subdirectly irreducible Abelian groups. For this example, some familiarity with Abelian group theory is assumed. In particular, the fact that every Abelian group is a module over the ring Z of integers with the definition $nx = x + x + \cdots + x$ (n-summands) for $n > 0$, $0x = 0$, $nx = (-n)x$ for $n < 0$ will be used. As in section 1.4, $[x]$ denotes the subgroup generated by x. It is easily seen that $[x] = \{nx \mid n \in Z\}$. A group $\mathbf{A}$ is called cyclic if $A = [x]$ for some $x \in A$. The order of an element $x \in A$ is defined to be $0(x) = |[x]|$. If $0(x)$ is finite and n is any integer, then $0(nx) = 0(x)/(n, 0(x))$ (where $(n, 0(x))$ is the greatest common divisor of n and $0(x)$). Of course, if $0(x)$ is infinite, then $0(nx)$ is infinite for all nonzero integers n.

It is well known that every congruence relation on an Abelian group is determined by a subgroup. In fact, it is easy to show that

$$\gamma: B \to \{\langle x, y\rangle \in A \times A \mid x - y \in B\}, \quad \text{and} \quad \sigma: \Gamma \to \Gamma^{-1}(0)$$

are order preserving correspondences between $\Sigma(\mathbf{A})$ and $\Theta(\mathbf{A})$ such that $\sigma\gamma(B) = B$ for all $B \in \Sigma(\mathbf{A})$. Therefore, $\Sigma(\mathbf{A})$ and $\Theta(\mathbf{A})$ are isomorphic as lattices.

We will now prove that the subdirectly irreducible Abelian groups are exactly the groups that are cyclic of prime power order, and the quasi-cyclic groups, that is, the groups that contain a system of generators $x_0, x_1, x_2, \cdots$ satisfying the relations $x_0 \neq 0, px_0 = 0, px_1 = x_0, px_2 = x_1, \cdots$, where p is any rational prime. Such a quasi-cyclic group is easily shown to be isomorphic to the multiplicative group of all p^kth roots of unity, $k = 1, 2, 3, \cdots$.

Assume that $A = [x]$, where x has order p^k (p is any prime). Let $M = [p^{k-1}x]$. Then $M \supset \{0\}$. If $y \neq 0$ in A, then $y = mp^s x$, where $s < k$ and $(m, p) = 1$. Hence, integers a and b exist satisfying $ap + bm = 1$. Consequently, $p^{k-1}x = bmp^{k-1}x + ap^k x = bp^{k-s-1}y \in [y]$. This shows that every nonzero subgroup B of $\mathbf{A}$ contains M. Thus, $\{0\}$ is meet irreducible in $\Sigma(\mathbf{A})$. Consequently, I_A is meet irreducible in $\Theta(\mathbf{A})$, so that $\mathbf{A}$ is subdirectly irreducible.

Assume that $A = [x_0, x_1, x_2, \cdots]$, where $x_0 \neq 0$, $px_0 = 0$, $px_1 = x_0$, $px_2 = x_1, \cdots$. Let $M = [x_0]$. It is clear that if $y \neq 0$ in A, then it is possible to write $y = a_0x_{i_0} + a_1x_{i_1} + \cdots + a_kx_{i_k}$, where $0 \leq i_0 < i_1 < \cdots < i_k$, and $a_0, a_1, \cdots, a_k$ are integers not divisible by p. Let b be an integer such that $a_kb \equiv 1 \pmod p$. Then $bp^{i_k}y = ba_kx_0 = x_0$. Hence $M \subseteq [y]$. This shows that every nonzero subgroup B of **A** contains M. Thus, as before, **A** is subdirectly irreducible.

We wish now to prove the converse: every subdirectly irreducible group is either cyclic of prime power order, or else quasi-cyclic. Assume that **A** is subdirectly irreducible. Since $\Sigma(\mathbf{A})$ is isomorphic to $\Theta(\mathbf{A})$, it follows from 1.6 that $\{0\}$ is meet irreducible. Hence, there is a subgroup M of **A** such that $\{0\} \subset M \subseteq B$ for all $B \in \Sigma(\mathbf{A})$ different from $\{0\}$. In particular, there is no subgroup C with $\{0\} \subset C \subset M$. Therefore, if $x \neq 0$ is in M, then $[x] = M$. Moreover, it is possible to choose an x having prime order p. In fact, if $2x \neq 0$, then $[2x] = [x]$, so that $x = 2kx$ for some $k \neq 1$. Hence, every x in M has finite order, and by choosing a nonzero element of minimal order, we obtain an element of prime order p. It is then clear that every nonzero element of M has order p. Henceforth, let $M = [x]$, $x \neq 0$, $px = 0$.

Let $y \in A$, $y \neq 0$. Then $[y] \supseteq [x]$. Hence, there exists an integer n such that $ny = x$. Consequently, $npy = 0$. Therefore $0(y)$ is finite and $0(y)$ divides np. If $(0(y), p) = 1$, then $0(y)$ divides n, so that $x = ny = 0$. This contradiction shows that p divides $0(y)$. To summarize, we have shown that every nonzero element of A has finite order divisible by p. Suppose that $0(y) = p^rm$ where $(p, m) = 1$. Then $0(p^ry) = m$ and p does not divide m. Hence, $p^ry = 0$ and $m = 1$. This shows that the order of every element of A is a power of p.

Next, we prove that if y and z have the same order p^r, then $[y] = [z]$. The proof is by induction on r. If $r = 0$, there is nothing to prove. Suppose that $r = 1$. Then $[y] \supseteq [x]$ and $|[y]| = p = |[x]|$ imply $[y] = [x]$. Similarly, $[z] = [x]$. Assume that $r > 1$ and the assertion holds for $r - 1$. Since $0(py) = 0(pz) = p^{r-1}$, the induction hypothesis yields $[py] = [pz]$. Thus, there is an integer n such that $py = npz$. Consequently, $p(y - nz) = 0$, so that either $0(y - nz) = p$, or else $0(y - nz) = 1$. In either case, $y - nz \in [y - nz] \subseteq [x] = [p^{r-1}z]$. Hence,

$$y = nz + p^{r-1}mz = (n + p^{r-1}m)z \in [z].$$

Therefore, $[y] \subseteq [z]$. Since $|[y]| = p^r = |[z]|$, we conclude that $[y] = [z]$. The induction is complete.

The characterization of subdirectly irreducible groups is obtained by considering two cases.

Case 1. The orders of the elements of A are bounded. Let $y \in A$ have maximal order p^s. Suppose that $z \in A$ and $0(z) = p^r \leq p^s$. Then $0(p^{s-r}y) = 0(z)$, so that $z \in [z] = [p^{s-r}y] \subseteq [y]$. This shows that $[y] = A$. Therefore **A** is cyclic.

Case 2. The orders of the elements of A are unbounded. In this case it is possible to choose $y_1, y_2, y_3, \cdots$ in A such that $0(y_n) = p^{n+1}$. Thus $0(py_n) = p^n$. Therefore, $[x] = [py_1] \subseteq [y_1]$ and $[y_n] = [py_{n+1}] \subseteq [y_{n+1}]$ for $n \geq 1$. Define $x_0 = x$. Suppose that x_m has been defined for $0 \leq m \leq n$ so that $px_m = x_{m-1}$ and $[x_m] = [y_m]$ for $m \geq 1$. Since $[py_{n+1}] = [y_n] = [x_n]$, there exists k such that $x_n = kpy_{n+1}$. Let $x_{n+1} = ky_{n+1}$. Then $px_{n+1} = x_n$, so that $0(x_{n+1}) = p^{n+1} = 0(y_{n+1})$. Thus $[x_{n+1}] = [y_{n+1}]$. This inductive construction yields elements $x_0, x_1, x_2, \cdots$ such that $B = [x_0, x_1, x_2, \cdots]$ is quasi-cyclic. If $z \in A$ and $0(z) = p^s$, then $z = 0 \in B$ if $s = 0$, and $z \in [z] = [x_{s-1}] \subseteq B$ if $s > 0$. Hence, $A = B$. That is, **A** is quasi-cyclic.

By Corollary 3.5 and the fact that every epimorphic image of an Abelian group is an Abelian group, it follows that every Abelian group is isomorphic to a subgroup of a direct product of groups which are cyclic of prime power order, or quasi-cyclic.

Problems

1. (a) Let $\mathfrak{A}$ be a class of partial algebras of type τ. Prove that a partial algebra $\mathbf{A}$ of type τ has a subdirect decomposition $\langle \varphi; \Pi_{i \in I}\mathbf{A}_i \rangle$ in which each $\mathbf{A}_i \in \mathfrak{A}$ if and only if $N_{\mathfrak{A}}(\mathbf{A}) = I_A$ (see Problem 12 in Chapter 1).

(b) Let $\mathfrak{A}$ be an abstract class of algebras of type τ. Prove that $\mathfrak{A}$ is closed under the formation of subdirect products (that is, any subdirect product of algebras in $\mathfrak{A}$ is in $\mathfrak{A}$) if and only if $\mathfrak{A}$ is residual (see Problem 11 in Chapter 1).

2. Let $\mathbf{A} = \langle A; F, G, H \rangle$, be the partial algebra of type $\{\langle 0, 1\rangle, \langle 1, 1\rangle, \langle 2, 1\rangle\}$ defined as follows:

(i) $A = \omega$.
(ii) $\mathscr{D}(F) = \omega$, $F(n) = n + 1$ for $n \in \omega$.
(iii) $\mathscr{D}(G) = \{n \in \omega \mid n > 0\}$, $G(n) = n + 1$ for $n > 0$.
(iv) $\mathscr{D}(H) = \{n \in \omega \mid n > 0\}$, $H(n) = n - 1$ for $n > 0$.

(a) Prove that every congruence relation on $\mathbf{A}$ is of the form $\Gamma_k = \{\langle m, n\rangle \in A \times A \mid m \equiv n \pmod{k}\}$ for some $k \geq 0$ in ω.

(b) Suppose that $\langle \varphi; \Pi_{i \in I}\mathbf{A}_i \rangle$ is a proper subdirect decomposition of $\mathbf{A}$, where $\Pi_{i \in I}\mathbf{A}_i = \langle \mathbf{X}_{i \in I}A_i; \bar{F}, \bar{G}, \bar{H} \rangle$. Show that $\bar{F}, \bar{G}, \bar{H}$ are operations, but $\bar{H}(\varphi(0)) \notin \varphi(A)$. Hence, $\varphi(A)$ does not determine a subalgebra of $\Pi_{i \in I}\mathbf{A}_i$.

(c) With the hypothesis of **(b)**, show that $\bar{G}(\varphi(A)) \subseteq \varphi(A)$, so that $\bar{G} \cap (\varphi(A))^2$ is an operation on $\varphi(A)$. Hence, $\mathbf{A}$ is not isomorphic to $(\Pi_{i \in I}\mathbf{A}_i) \restriction \varphi(A)$.

This example demonstrates the failure of 3.5 when the condition that $\mathbf{A}$ be an algebra is relaxed.

3. Let **A** be a finitary algebra.

(a) Show that there is an algebra **B** with unary operations such that $\Theta(\mathbf{A}) = \Theta(\mathbf{B})$. (*Hint:* let **B** be the algebra on the set A with all the unary operations obtained by varying only one component in the operations of **A**.)

(b) Show that there is a semigroup with operators **S** such that $\Theta(\mathbf{A}) = \Theta(\mathbf{S})$. (*Hint:* by **(a)** it can be assumed that $\mathbf{A} = \langle A; F_\xi \rangle_{\xi < \rho}$, with all F_ξ unary. Define $x \circ y = x$ for all x, y in A. Let $\mathbf{S} = \langle A; \circ, F_\xi \rangle_{\xi < \rho}$.)

4. Let **L** be a complete lattice. A *meet representation* of **L** is a pair $\langle X; G \rangle$ consisting of a set X and a mapping G of L to $\mathbf{P}(X)$ such that

(i) $a \leq b$ in **L** if and only if $G(a) \subseteq G(b)$.

(ii) $G(\bigwedge_{i \in I} a_i) = \bigcap_{i \in I} G(a_i)$.

(a) Show that if $\langle X; G \rangle$ is a meet representation of **L**, then for each $x \in X$, there exists a unique $e_x \in L$ such that

(iii) $x \in G(a)$ if and only if $a \geq e_x$.

Prove that for every $a \in L$, a is the least upper bound of $\{e_x \mid x \in G(a)\}$.

(b) Suppose that G is a mapping of L into $\mathbf{P}(X)$ satisfying (i) and such that for each $x \in X$, there is an element $e_x \in L$ for which (iii) holds. Prove that $\langle X; G \rangle$ is a meet representation of **L**.

(c) Let X be a dense subset of **L**, that is a set such that $a = \bigvee \{x \in X \mid x \leq a\}$ for all $a \in L$. For $a \in L$, define $G(a) = \{x \in X \mid x \leq a\}$. Show that $\langle X; G \rangle$ is a meet representation of **L** such that $e_x = x$ for all $x \in X$.

(d) Show that if **L** has a meet representation $\langle X; G \rangle$ such that for all $x \in X$, the element e_x [defined as in **(a)**] is compact, then **L** is compactly generated, and for any directed set $D \subseteq L$, $G(\bigvee D) = \bigcup \{G(d) \mid d \in D\}$.

(e) Use the results of **(c)**, **(d)**, and 2.9 to prove that if **L** is a compactly generated lattice and **L**$'$ is a sublattice of **L** such that if $X \subseteq L'$, then $\bigvee X \in L'$ and $\bigwedge X \in L'$, then **L**$'$ is compactly generated.

(f) Use the results of **(e)** and 3.2 to show that if $a < b$ in the compactly generated lattice **L**, then c and d exist in L such that $a \leq c < d \leq b$, and $\{x \in L \mid c < x < d\} = \varnothing$. A lattice with this property is called *weakly atomic.*

(g) Use the results of **(c)** and **(d)** above to prove that every compactly generated lattice B is *upper continuous:* if $D = \{b_i \mid i \in I\}$ is a directed subset of L and $a \in L$, then $a \wedge \bigvee_{i \in I} b_i = \bigvee_{i \in I} a \wedge b_i$.

5. Let $\mathscr{L} \subseteq \mathbf{P}(X)$ be closed under intersections and closed under unions of directed sets, as in 2.9. For $Z \subseteq X$, define $[Z] = \bigcap \{A \in \mathscr{L} \mid Z \subseteq A\}$.

(a) Prove that $[Z_1 \cup Z_2] = [Z_1] \vee [Z_2]$ in $\mathscr{L}$.

(b) Prove that $c \in \mathscr{L}$ is compact if and only if $c = [Z]$ for some finite set $Z \subseteq X$.

(c) Use parts **(a)** and **(b)** of this problem, together with **4(d)**, to show that the join of two compact elements in a compactly generated lattice is compact.

(d) Prove that if $A \in \mathscr{L}$, then $A = \bigcup \{[X] \mid X \subseteq A, X \text{ finite}\}$.

(e) Prove that there is a finitary algebra $\mathbf{X} = \langle X; F_\xi \rangle_{\xi < \rho}$ such that $\Sigma(\mathbf{X}) = \mathscr{L}$. (*Hint:* let $\{F_\xi \mid \xi < \rho\}$ be the set of all finitary operations F such that $F(\mathbf{x}) \in [\{x(\eta) \mid \eta < \tau\}]$ if $\mathbf{x} \in \mathscr{D}(F)$.)

(f) Show that the following conditions are equivalent for a lattice $\mathbf{L}$:

(i) $\mathbf{L} \cong \Sigma(\mathbf{A})$ for some finitary algebra $\mathbf{A}$.
(ii) $\mathbf{L} \cong \Sigma(\mathbf{A})$ for some finitary partial algebra $\mathbf{A}$.
(iii) $\mathbf{L}$ is compactly generated.

6. Prove that every complete lattice is isomorphic to $\Sigma(\mathbf{A})$ for some algebra $\mathbf{A}$.

7. Let $\mathbf{L}$ be a compactly generated lattice with smallest element o.

(a) Let C be the set of all compact elements of $\mathbf{L}$. Define

$$G(a) = \{\langle x, y \rangle \in C \times C \mid x = y \text{ or } x \vee y \leq a\}.$$

Prove that $G(a) \in E(C)$. Show that $\langle C \times C; G \rangle$ is a representation of $\mathbf{L}$ such that $G(o) = I_C$, and $\mathscr{L} = G(L)$ is closed under intersections and closed under unions of directed sets.

(b) Let $\mathscr{L} \subseteq E(X)$ be such that $\mathscr{L}$ is closed under intersections and closed under unions of directed sets. Assume that $I_X \in \mathscr{L}$. Prove that there is a finitary partial algebra $\mathbf{X} = \langle X; F_\xi \rangle_{\xi < \rho}$ such that $\Theta(\mathbf{X}) = \mathscr{L}$. (*Hint:* let $\{F_\xi \mid \xi < \rho\}$ be a well ordering of the set of all partial operations F such that $\mathscr{D}(F) = \{\mathbf{x}, \mathbf{y}\}$, $\mathbf{x}$ and $\mathbf{y}$ in X^n, and

$$\langle F(\mathbf{x}), F(\mathbf{y}) \rangle \in [\{\langle \mathbf{x}(0), \mathbf{y}(0) \rangle, \langle \mathbf{x}(1), \mathbf{y}(1) \rangle, \cdots, \langle \mathbf{x}(n-1), \mathbf{y}(n-1) \rangle\}],$$

where the symbolism $[\cdots]$ has the same meaning as in Problem 5.)

8. Let $\mathbf{L}$ be a complete lattice. Prove that $\mathbf{L}$ is isomorphic to $\Theta(\mathbf{A})$ for some partial algebra $\mathbf{A}$.

9. Let $\mathbf{V}$ be a two-dimensional vector space over a prime field $\mathbf{F}$ with more than two elements. That is, $\mathbf{F}$ is either the rational field, or the field of integers modulo p, where p is a prime greater than 2. Thus, V consists of all pairs $\langle x, y \rangle$, where x and y belong to F. For a point $p = \langle x, y \rangle$, it is convenient to write p_x for x and p_y for y, the components of p. Let $\mathscr{L}$ consist of the following subsets of $V \times V$:

$$\begin{aligned} I_V &= \{\langle p, p \rangle \mid p \in V\}, \\ \Gamma_X &= \{\langle p, q \rangle \mid p_x = q_x\}, \\ \Gamma_Y &= \{\langle p, q \rangle \mid p_y = q_y\}, \\ \Gamma_D &= \{\langle p, q \rangle \mid p_x - q_x = p_y - q_y\}, \\ U_V &= V \times V. \end{aligned}$$

Show that $\mathscr{L}$ satisfies the conditions of **7(b)**, but there is no abstract algebra $\mathbf{A}$ such that $\Theta(\mathbf{A}) = \mathscr{L}$.

10. Let $\mathbf{B} = \langle B; \bigvee_{n<\omega}, ' \rangle$ be the ω-complete Boolean algebra of measurable subsets of the real numbers, modulo sets of measure zero. The ω-ary operation $\bigvee_{n<\omega}$ is the least upper bound on countable sets, and $(')$ is complementation. Let o denote the least element of B.

(a) Prove that the correspondence $\Gamma \to \{a \in B \mid \langle a, o \rangle \in \Gamma\}$ is a lattice isomorphism between $\Theta(\mathbf{B})$ and the lattice $\mathbf{J}(\mathbf{B})$ of all σ-ideals of $\mathbf{B}$, that is, those sets $J \subseteq B$ such that $a \leq b \in J$ implies $a \in J$ and $\{a_n \mid n < \omega\} \subseteq J$ implies $\bigvee_{n<\omega} a_n \in J$.

(b) Suppose that J is a compact element of $\mathbf{J}(\mathbf{B})$. Prove that there exists $a \in B$ such that $J = \{b \in B \mid b \leq a\}$. (*Hint:* use Problem 5(b).)

(c) Show that if $a > o$ in $\mathbf{B}$, there exists a sequence $\{a_n \mid n < \omega\}$ such that $a_n \wedge a_m = o$ for $n \neq m$ and $\bigvee_{n<\omega} a_n = a$.

(d) Deduce from (a), (b), and (c) that there are no nonzero compact elements in $\Theta(\mathbf{B})$.

11. A partial algebra $\mathbf{A}$ is called *simple* if $|A| > 1$ and $\Theta(\mathbf{A}) = \{I_A, U_A\}$. A partial algebra $\mathbf{A}$ is called *semisimple* if there is a decomposition of $\mathbf{A}$ as a subdirect product of simple algebras. Note that every simple algebra is subdirectly irreducible.

(a) Show that $\mathbf{A}$ is semisimple if and only if I_A is an intersection of dual atoms in $\Theta(\mathbf{A})$. (If $\mathbf{L}$ is a lattice with greatest element u, an element $d \in L$ is called a *dual atom* if $d < u$ and $\{x \in L \mid d < x < u\} = \varnothing$.)

An algebra $\mathbf{A}$ is called *hereditarily semisimple* if every epimorphic image of $\mathbf{A}$ is semisimple.

(b) Suppose that $\mathbf{A}$ is a finitary abstract algebra. Prove that the following conditions are equivalent:

(i) $\mathbf{A}$ is hereditarily semisimple;
(ii) every element of $\Theta(\mathbf{A})$ is an intersection of dual atoms;
(iii) every meet irreducible element of $\Theta(\mathbf{A})$ is a dual atom.

12. Let $\mathbf{S} = \langle S; \vee \rangle$ be a *semilattice*, that is, an algebra with a single binary operation $\vee$ satisfying

$$a \vee (b \vee c) = (a \vee b) \vee c, \qquad a \vee b = b \vee a, \quad \text{and} \quad a \vee a = a$$

for all a, b, and c in S.

(a) Define $x \leq y$ in S if $x \vee y = y$. Prove that $\langle S; \leq \rangle$ is a partially ordered set such that $x \vee y$ is the least upper bound of x and y for all x, y in S.

(b) For $a \in S$, define $\Gamma_a = \{\langle x, y \rangle \in S \times S \mid x \leq a \text{ and } y \leq a, \text{ or } x \not\leq a$ and $y \not\leq a\}$. Show that Γ_a is a congruence relation on $\mathbf{S}$ such that either $\Gamma_a = S \times S$, or $\mathbf{S}/\Gamma_a$ is a two-element semilattice.

(c) Prove that every semilattice is hereditarily semisimple and that every subdirectly irreducible semilattice has two elements and is simple.

(d) Prove that every semilattice is isomorphic to a sub-semilattice of $\langle \mathbf{P}(X); \cup \rangle$ for some set X.

13. Let $\mathbf{L} = \langle L; \vee, \wedge \rangle$ be a distributive lattice (see 4.2).

(a) For $a \in L$, define $\Gamma_a = \{\langle x, y \rangle \in L \times L \mid x \wedge a = y \wedge a\}$, $\Delta_a = \{\langle x, y \rangle \in L \times L \mid x \vee a = y \vee a\}$. Prove that Γ_a and Δ_a are congruence relations on $\mathbf{L}$.

(b) Show that every simple distributive lattice contains exactly two elements.

(c) Prove that every distributive lattice is hereditarily semisimple. (*Hint*: suppose that $\mathbf{L}$ is subdirectly irreducible, and that $\Gamma_0 \supset I_L$, $\Gamma_0 \in \Theta(\mathbf{L})$ is such that $\Gamma_0 \subseteq \Gamma$ for all $\Gamma \neq I_L$ in $\Theta(\mathbf{L})$. Show that there exists $\langle a, b \rangle \in \Gamma_0$ with $a < b$. Prove that $\Delta_a = \Gamma_b = I_A$, and deduce that $\Theta(\mathbf{L}) = \{I_L, \Gamma_0 = U_L\}$.)

(d) Prove that every distributive lattice is isomorphic to a sublattice of $\langle \mathbf{P}(X); \cup, \cap \rangle$ for some set X.

14. A *lattice-ordered group* or *l*-group is an abstract algebra

$$\mathbf{G} = \langle G; +, 0, -, \vee, \wedge \rangle$$

such that

(i) $\langle G; +, 0, - \rangle$ is a group (not necessarily commutative‡),
(ii) $\langle G; \vee, \wedge \rangle$ is a lattice, and
(iii) $x \to a + x + b$ is a lattice automorphism for each a and b in G.

As usual, denote the partial ordering on G, which is derived from $\vee$ and $\wedge$ by $\leqq$.

The purpose of this problem is to develop the theory of *l*-groups to the point of proving a fundamental representation theorem.

(a) Prove that the class $\mathfrak{G}$ of all *l*-groups satisfies $\mathscr{Q}\mathfrak{G} = \mathscr{S}\mathfrak{G} = \mathscr{P}\mathfrak{G} = \mathfrak{G}$.

(b) Let $\mathbf{L}$ be a totally ordered set. Let $A(\mathbf{L})$ be the set of all order isomorphisms of $\mathbf{L}$, that is, one-to-one mappings α of L onto itself such that $x < y$ in $\mathbf{L}$ implies $\alpha(x) < \alpha(y)$. Define: $(\alpha + \beta)(x) = \alpha(\beta(x))$, $(-\alpha)(x) = \alpha^{-1}(x)$, $0(x) = x$, $(\alpha \vee \beta)(x) = \max \{\alpha(x), \beta(x)\}$, $(\alpha \wedge \beta)(x) = \min \{\alpha(x), \beta(x)\}$. Show that $\mathbf{A(L)} = \langle A(\mathbf{L}); +, 0, -, \vee, \wedge \rangle$ is an *l*-group. Find an $\mathbf{L}$ for which $\mathbf{A(L)}$ is not commutative.

In the remainder of this problem, $\mathbf{G}$ denotes a fixed *l*-group.

(c) Show that the following identities hold for the elements of G:

(i) $a + (b \vee c) = (a + b) \vee (a + c)$, $(b \vee c) + a = (b + a) \vee (c + a)$,
(ii) $a + (b \wedge c) = (a + b) \wedge (a + c)$, $(b \wedge c) + a = (b + a) \wedge (c + a)$,
(iii) $-(a \vee b) = (-a) \wedge (-b)$, $-(a \wedge b) = (-a) \vee (-b)$.

(d) Show that if $a \geqq 0$, $b \geqq 0$, and $c \geqq 0$, then $(a \wedge b) + (a \wedge c) \geqq a \wedge (b + c)$.

‡ The theory of *l*-groups is one of the few subjects in which $+$ is used to denote a non-commutative binary operation. This convention makes it possible to apply the theory of *l*-groups to the study of lattice-ordered rings without any change of notation.

(e) Prove by induction on n that $n(a \wedge 0) = na \wedge (n-1)a \wedge \cdots \wedge a \wedge 0$. Use this to show that if $na \geqq 0$ for $n \geqq 1$, then $a \geqq 0$.

(f) Show that for every $a \in G$, $2(a \vee -a) \geqq 0$, so that $a \vee -a \geqq 0$. For $a \in G$, define $|a| = a \vee -a$.

(g) Prove that for all a and b:

(i) $|-a| = |a|$,
(ii) $-|a| \leqq a \leqq |a|$;
(iii) $|a| \geqq 0$, and $|a| = 0$ only if $a = 0$;
(iv) $|a + b| \leqq |a| + |b| + |a|$;
(v) if G is commutative, then $|a + b| \leqq |a| + |b|$.

A subgroup H of G is called *l-convex* if

$$a \in G, \qquad b \in H, \qquad |a| \leqq |b| \quad \text{implies} \quad a \in H.$$

(h) Prove that the set $\mathscr{L}$ of all l-convex subgroups of $\mathbf{G}$ satisfies the conditions (i) and (ii) of Proposition 2.9.

(i) Let $g \neq 0$ in G. Prove that there is an $H \in \mathscr{L}$ which is maximal with the property that $g \notin H$. Show that for such an H there is a smallest element in $\{K \in \mathscr{L} \mid K \supset H\}$.

(j) Let $H \in \mathscr{L}$, and $a \geqq 0$. Define $K = \{x \in G \mid a \wedge |x| \in H\}$. Show that $K \in \mathscr{L}$ and $H \subseteq K$.

(k) Let $H \in \mathscr{L}$, and $a \wedge b = 0$ in $\mathbf{G}$. Define K as in (j), and $K' = \bigcap_{x \in K}\{y \in G \mid |x| \wedge |y| \in H\}$. Show that $K' \in \mathscr{L}$, $b \in K$, $a \in K'$, and $K \cap K' = H$.

(l) Let $g \neq 0$. Suppose that H is an l-convex subgroup of G that is maximal with respect to the property $g \notin H$. Let $L(H) = \{x + H \mid x \in G\}$ be the set of all left cosets of H. Define $a + H \leqq b + H$ if there exists $h \in H$ such that $a + h \leqq b$. Prove that $\leqq$ is a total ordering of $L(H)$ such that $(a \wedge b) + H = \min\{a + H, b + H\}$ for all $a \in G$ and $b \in G$.

(m) With the same assumptions as in (l), define $\hat{a}(x + H) = (a + x) + H$. Prove that $\hat{a}$ is a well-defined order automorphism of $\mathbf{L}(H) = \langle L(H), \leq\rangle$, and the mapping $a \to \hat{a}$ is a homomorphism of $\mathbf{G}$ (as an l-group) onto a transitive sub-l-group of $\mathbf{A}(\mathbf{L}(H))$. (Transitive means that if $x + H$ and $y + H$ are two elements of $L(H)$, then $\hat{a}(x + H) = y + H$ for some $a \in G$.) Finally, show that $\hat{g}$ is not the zero element of $\mathbf{A}(\mathbf{L}(H))$.

(n) Let $\mathbf{G}$ be subdirectly irreducible. Prove that there exists $g \in G$ such that every homomorphism φ of $\mathbf{G}$ that is not a monomorphism satisfies $\varphi(g) = 0$. Deduce that G is isomorphic to a transitive sub-l-group of $\mathbf{A}(\mathbf{L}(H))$ for a suitable l-convex subgroup H of $\mathbf{G}$.

(o) Let $\{\mathbf{L}_\xi \mid \xi < \lambda\}$ be a well-ordered sequence of totally ordered sets. Let $L = \bigcup_{\xi<\lambda}(L_\xi \times \{\xi\})$, and define $\langle x, \xi\rangle \leq \langle y, \eta\rangle$ if $\xi < \eta$, or $\xi = \eta$ and $x \leq y$ in $\mathbf{L}_\xi$. Show that $\mathbf{L} = \langle L, \leq\rangle$ is totally ordered. Prove that there is an l-group monomorphism of $\Pi_{\xi<\lambda}\mathbf{A}(\mathbf{L}_\xi)$ into $\mathbf{A}(\mathbf{L})$. Use this result and Theorem 3.4 to prove the following:

Theorem. (Holland [26]) Every lattice-ordered group is isomorphic to a sub-l-group of $\mathbf{A}(\mathbf{L})$ for a suitable totally ordered set $\mathbf{L}$.

(p) Use the result of **(o)** to show that every l-group is distributive (as a lattice). (Remark: This fact can be proved directly without much trouble, but it is interesting to note that the proof of the representation theorem **(o)** does not use distributivity, and that distributivity is an easy consequence of the representation.)

15. A *function ring* or f-ring is an abstract algebra $\mathbf{R} = \langle R; +, 0, -, \cdot, \vee, \wedge\rangle$ such that

(i) $\langle R; +, 0, -, \vee, \wedge\rangle$ is a commutative l-group,
(ii) $\langle R; +, 0, -, \cdot\rangle$ is an associative ring, and
(iii) $a \wedge b = 0$, $c \geqq 0$ in $\mathbf{R}$ imply $ac \wedge b = ca \wedge b = 0$.

(a) Prove that the class $\mathfrak{R}$ of all f-rings satisfies $\mathscr{Q}\mathfrak{R} = \mathscr{S}\mathfrak{R} = \mathscr{P}\mathfrak{R} = \mathfrak{R}$. Show that $\mathfrak{R}$ contains the class of all ordered rings, that is, rings that have a total ordering satisfying $a \leqq b$ implies $a + c \leqq b + c$ for all c, and $a \geqq 0$, $b \geqq 0$ implies $ab \geqq 0$.

In the remainder of this problem, $\mathbf{R}$ denotes a fixed f-ring. Our objective is to prove a representation theorem for f-rings.

(b) Show that the following identities hold for the elements of R:

(i) $a + b = (a \vee b) + (a \wedge b)$,
(ii) $a \geqq 0$ implies $a(b \vee c) = (ab) \vee (ac)$, $a(b \wedge c) = (ab) \wedge (ac)$, $(b \vee c)a = (ba) \vee (ca)$, $(b \wedge c)a = (ba) \wedge (ca)$,
(iii) $a \geqq 0$ and $b \geqq 0$ implies $ab \geqq 0$,
(iv) $a \wedge b = 0$ implies $ab = 0$,
(v) $a^2 \geqq 0$.

For any element $a \in R$, define $a^+ = a \vee 0$, $a^- = a \wedge 0$.

(c) Prove that for every $a \in R$, $a^+ + a^- = a$ and $a^+ - a^- = |a|$. Use this result to prove the following:

(i) $|a - b| = (a \vee b) - (a \wedge b)$
(ii) $|(a \vee c) - (b \vee c)| \leqq |a - b|$, $|(a \wedge c) - (b \wedge c)| \leqq |a - b|$
(*Hint*: use (i) and the result stated in part **(p)** of Problem 14 above.)
(iii) $|a \cdot b| \leqq |a| \cdot |b|$

An *l-ideal* of $\mathbf{R}$ is a nonempty set $J \subseteq R$ such that J is a two-sided ring ideal that satisfies

$$|a| \leqq |b| \quad \text{and} \quad b \in J \quad \text{implies} \quad a \in J.$$

(d) Show that the set $\mathscr{L}$ of all l-ideals of $\mathbf{R}$ satisfies the conditions (i) and (ii) of Proposition 2.9, so that $\langle \mathscr{L}; \subseteq\rangle$ is a complete lattice.

(e) Prove that $\Gamma \to \bar{\Gamma}^{-1}(0)$, $J \to \{\langle a, b\rangle \mid a - b \in J\}$ are inverse lattice isomorphisms between $\Theta(\mathbf{R})$ and $\mathscr{L}$.

(f) Suppose that $a \wedge b = 0$ in $\mathbf{R}$. Define $J = \{c \in R \mid a \wedge |c| = 0\}$ and $K = \bigcap_{x \in J}\{y \in R \mid |x| \wedge |y| = 0\}$. Prove that J and K are l-ideals such that $b \in J$, $a \in K$, and $J \cap K = \{0\}$.

(g) Prove that every subdirectly irreducible f-ring is totally ordered. Hence, deduce:

Theorem. (Birkhoff and Pierce [5]) Every f-ring is isomorphic to a subdirect product of ordered rings.

16. Let $\mathbf{R} = \langle R; +, 0, -, \cdot\rangle$ be a commutative ring.

(a) Prove that there is a lattice isomorphism from the lattice of ideals of $\mathbf{R}$ to $\Theta(\mathbf{R})$ given by

$$J \to \Gamma_J = \{\langle x, y\rangle \in R \times R \mid x - y \in J\}.$$

(b) Show that if $\mathbf{R}$ is a subdirectly irreducible commutative ring with no nonzero nilpotent elements, then $\mathbf{R}$ is a field. (*Hint*: let J be the smallest nonzero ideal of $\mathbf{R}$; show that if $x \in J$, then $x \cdot z \neq 0$ for all $z \in R$; deduce that $x \cdot J = J$ for $x \in J$, and consequently there is an $e \in J$ such that $x \cdot e = x$; prove that $e \cdot z = z$ for all $z \in R$, so that $J = R$, and $\mathbf{R}$ is a field.)

(c) Prove that every commutative regular ring is isomorphic to a subdirect product of fields. (A ring is called regular if for all x there exists y such that $xyx = x$.)

17. Let $\mathbf{B} = \langle B; \vee, \wedge, o, u, '\rangle$ be a Boolean algebra. An ideal of $\mathbf{B}$ is a nonempty set $J \subseteq B$ such that

(i) if $x \in J$ and $y \in J$, then $x \vee y \in J$;
(ii) if $x \in J$ and $z \leq x$, then $z \in J$.

(a) Prove that the correspondence

$$\Gamma \to J(\Gamma) = \{x \in B \mid \langle x, o\rangle \in \Gamma\}$$

is an isomorphism of $\Theta(\mathbf{B})$ to the lattice $\mathbf{J(B)}$ of all ideals of $\mathbf{B}$.

(b) Show that if J and K are ideals of $\mathbf{B}$, then $J \vee K = \{x \vee y \mid x \in J, y \in K\}$ is the least upper bound of $\{J, K\}$ in $\mathbf{J(B)}$. Use this result to show that $\mathbf{J(B)}$ [and hence $\Theta(\mathbf{B})$] is a distributive lattice.

18. Let $\mathbf{L}$ be a compactly generated, distributive lattice with least element o and greatest element u. Let M be the set of all meet irreducible elements of $\mathbf{L}$. For $a \in L$, define $h(a) = \{m \in M \mid m \geq a\}$. For $X \subseteq M$, let $X^- = h(\bigwedge X)$.

(a) Prove that $X \to X^-$ is a topological closure operation, that is, $X^- \supseteq X$, $X^{--} = X^-$, and $(X \cup Y)^- = X^- \cup Y^-$.

(b) Prove that the mapping $a \to h(a)$ is a one-to-one, order-reversing correspondence between $\mathbf{L}$ and the set of all closed subsets of the space M with this closure operation.

(c) Prove that an element $a \in L$ is compact as a lattice element if and only if the set $M - h(a)$ is compact as a subspace of M.

(d) Prove that with this closure operation, M is a T_1 space if and only if every meet irreducible element of $\mathbf{L}$ is a dual atom (see Problem 11).

(e) Prove that with this closure operation, M is a Hausdorff space if and only if every compact element of $\mathbf{L}$ is complemented, that is, if c is compact, there exists $a \in L$ such that $c \wedge a = o$, $c \vee a = u$. Show that in this case M is zero dimensional, that is, M has a basis of neighborhoods that are open and closed.

(f) Prove that with this closure operation, M is a compact Hausdorff space if and only if the set B of all compact elements of $\mathbf{L}$ determines a sublattice of $\mathbf{L}$ such that $\mathbf{B} = \mathbf{L} \restriction B$ is a Boolean algebra. Show that in this case $\mathbf{L}$ is isomorphic to the lattice $\Theta(\mathbf{B})$.

19. Let $\mathbf{A}$ and $\mathbf{B}$ be similar partial algebras. Suppose that $\varphi \in \mathrm{Hom}\,(\mathbf{A}, \mathbf{B})$. For $\Gamma \in \Theta(\mathbf{B})$, define

$$\gamma_\varphi(\Gamma) = \{\langle a, b\rangle \in A \times A \mid \langle \varphi(a), \varphi(b)\rangle \in \Gamma\}.$$

(a) Show that γ_φ is an order-preserving mapping of $\Theta(\mathbf{B})$ to $\Theta(\mathbf{A})$.

(b) Prove that if φ is an epimorphism, then γ_φ is one-to-one. Identify the image of γ_φ.

(c) Prove that if φ is an epimorphism, then $\mathbf{A}/\gamma_\varphi(\Gamma)$ is isomorphic to $\mathbf{B}/\Gamma$. From this, it follows that if Γ is meet irreducible, then so is $\gamma_\varphi(\Gamma)$.

(d) Show that if $\Theta(\mathbf{A})$ and $\Theta(\mathbf{B})$ are distributive and φ is an epimorphism, then γ_φ is a continuous mapping of the space of meet irreducible elements of $\Theta(\mathbf{B})$ into the space of meet irreducible elements of $\Theta(\mathbf{A})$ (see Problem 18).

Notes to Chapter 2

The fundamental subdirect decomposition theorem 3.4 is due to Birkhoff [2]. The approach to this result by the path of compactly generated lattices is implicit in the presentation given in Birkhoff's book [3]. A deep study of compactly generated lattices has recently been carried out by Crawley and Dilworth (see [15] and [16]). Grätzer and Schmidt have studied the lattice $\Theta(\mathbf{A})$ in detail. The results sketched in Problems 3 and 7 are taken from their works [22] and [23]. The main result, which they prove in [23], is the difficult theorem that every compactly generated lattice is isomorphic to $\Theta(\mathbf{A})$ for some finitary algebra $\mathbf{A}$. (See also Grätzer's book [21].) It would be interesting to characterize those sublattices Θ of $E(A)$ such that there is a (finitary) abstract algebra $\mathbf{A} = \langle A; F_\xi\rangle_{\xi<\rho}$ satisfying $\Theta(\mathbf{A}) = \Theta$. Problem 9 shows that not every sublattice of $E(A)$ has this property. Notice that the corresponding problem for $\Sigma(\mathbf{A})$ has a very simple answer (Problem 5(e)) due to Birkhoff and Frink [4].

[3]

Direct Decompositions

I. Interdirect Decompositions

It has been observed that a set of relational systems can have many different subdirect products. However, there is one and only one direct product of a given set of similar systems. Consequently, if a relational system $\mathbf{A}$ can be expressed as a direct product of a set $\{\mathbf{A}_i \mid i \in I\}$ of relational systems, then $\mathbf{A}$ is completely determined by the systems $\mathbf{A}_i$. For this reason, the study of direct decompositions is in many ways more fundamental than investigations of subdirect decompositions.

Let $\mathbf{A}$ be a relational system. Let $\{\mathbf{A}_i \mid i \in I\}$ be a set of relational systems which are similar to $\mathbf{A}$. A *decomposition of* $\mathbf{A}$ *as a direct product of* $\{\mathbf{A}_i \mid i \in I\}$ is an isomorphism φ of $\mathbf{A}$ to the direct product $\Pi_{i\in I}\mathbf{A}_i$. The decomposition is called *proper* if $\langle \varphi; \Pi_{i\in I}\mathbf{A}_i\rangle$ is a proper subdirect decomposition of $\mathbf{A}$. A relational system is called indecomposable if it admits no decomposition as a proper direct product.

It is easy to see that if $\mathbf{A}$ is a relational system with more than one element and no empty relations, then a decomposition as a direct product of $\{\mathbf{A}_i \mid i \in I\}$ is proper if and only if $|I| > 1$ and at least two of the A_i have more than one element.

As in the case of subdirect decompositions, we would like to relate direct decompositions to the lattice of congruence relations. This is possible, but the correspondence is not very elegant unless I is finite. It is possible to define another concept intermediate between direct decompositions and subdirect decompositions, which coincides with the concept of a direct decomposition if I is finite. Moreover, such decompositions can be characterized very nicely in terms of $\Theta(\mathbf{A})$.

1.1 DEFINITION. Let $\{\mathbf{A}_i \mid i \in I\}$ be a set of relational systems of type τ. An *interdirect product* of $\{\mathbf{A}_i \mid i \in I\}$ is a subdirect product

$$\mathbf{B} = (\Pi_{i\in I}\mathbf{A}_i) \restriction B,$$

with the property that if $a \in \mathbf{X}_{i\in I}A_i$ and $b \in B$ satisfy $|\{i \in I \mid \pi_i(a) \neq \pi_i(b)\}| < \aleph_0$, then $a \in B$.

It is clear that the direct product is an interdirect product. However, an infinite set of relational systems generally has many different interdirect products. Suppose for example that $\{\mathbf{A}_i \mid i \in I\}$ is a set of Abelian groups. The *direct sum* of $\{\mathbf{A}_i \mid i \in I\}$ is defined to be the subalgebra $\Sigma_{i\in I}\mathbf{A}_i$ of the direct product $\Pi_{i\in I}\mathbf{A}_i$ determined by the set $B = \{a \in \mathbf{X}_{i\in I}A_i \mid a(i) = 0$, for all but finitely many $i\}$. Then $\Sigma_{i\in I}\mathbf{A}_i$ is an interdirect product of $\{\mathbf{A}_i \mid i \in I\}$. Moreover, it is easy to see in this case that the interdirect products of $\{\mathbf{A}_i \mid i \in I\}$ are exactly the subalgebras of $\Pi_{i\in I}\mathbf{A}_i$ which contain $\Sigma_{i\in I}\mathbf{A}_i$ as a subalgebra.

1.2 DEFINITION. A decomposition of a relational system $\mathbf{A}$ as an *interdirect product* of a set $\{\mathbf{A}_i \mid i \in I\}$ of relational systems similar to $\mathbf{A}$ is a monomorphism φ of $\mathbf{A}$ to $\Pi_{i\in I}\mathbf{A}_i$ such that

(i) $\langle \varphi; \Pi_{i\in I}\mathbf{A}_i\rangle$ is a subdirect decomposition of $\mathbf{A}$, and
(ii) $(\Pi_{i\in I}\mathbf{A}_i) \restriction \varphi(A)$ is an interdirect product of $\{\mathbf{A}_i \mid i \in I\}$.

As in the case of subdirect decompositions, it will be convenient to refer to the *interdirect* decomposition $\langle \varphi; \Pi_{i\in I}\mathbf{A}_i\rangle$.

If φ is an interdirect decomposition of $\mathbf{A}$, then φ is not necessarily an isomorphism of $\mathbf{A}$ to $(\Pi_{i\in I}\mathbf{A}_i) \restriction \varphi(A)$. Moreover, $\varphi(A)$ may not determine a subalgebra of $\Pi_{i\in I}A_i$ (see Problem 1). However, by 2.1.2, both of these conditions are satisfied if $\mathbf{A}$ is an algebra.

The relation between interdirect products and congruence relations is stated in terms of the composition of binary relations, which we now define.

1.3 DEFINITION. Let Γ and Δ be binary relations on a set X. The *composition** of Γ and Δ is the binary relation

$$\Gamma \circ \Delta = \{\langle x, z\rangle \in X \times X \mid \langle x, y\rangle \in \Gamma \quad \text{and} \quad \langle y, z\rangle \in \Delta \quad \text{for some} \quad y \in X\}.$$

1.4 PROPOSITION. Let $\langle \varphi; \Pi_{i\in I}\mathbf{A}_i\rangle$ be an interdirect decomposition of $\mathbf{A}$, where $\mathbf{A}$ and $\{\mathbf{A}_i \mid i \in I\}$ are similar partial algebras. Let Γ_i be the kernel of the epimorphism $\pi_i \circ \varphi$. Define

$$\Gamma^i = \textstyle\bigcap_{j\neq i}\Gamma_j.$$

Then

(i) $\bigcap_{i\in I}\Gamma_i = I_A$,
(ii) $\Gamma^i \circ \Gamma_i = U_A$ for all $i \in I$.

Proof. The assertion (i) is a consequence of the fact that $\langle \varphi; \Pi_{i\in I}\mathbf{A}_i\rangle$ is a subdirect decomposition of $\mathbf{A}$. To prove (ii), suppose that $\langle x, z\rangle \in U_A = A \times A$. Let $\mathbf{B} = (\Pi_{i\in I}\mathbf{A}_i) \restriction \varphi(A)$. By assumption, $\mathbf{B}$ is an interdirect product of $\{\mathbf{A}_i \mid i \in I\}$. Let a be the element of the product $\mathbf{X}_{i\in I}A_i$ such that $\pi_j(a) = \pi_j(\varphi(x))$ for all $j \neq i$, and $\pi_i(a) = \pi_i\varphi(z)$. Clearly, $\{j \in I \mid \pi_j(a) \neq \pi_j(\varphi(x))\}$ is finite. Therefore, by the definition of an interdirect product,

* The habit of writing a function symbol on the left of its argument leads to a definition for the composition of mappings, which is the reverse of the composition of relations defined here. In practice, this ambiguity causes very little confusion.

$a \in B$. That is, $a = \varphi(y)$ for some $y \in A$. Note that $\pi_j\varphi(x) = \pi_j\varphi(y)$ for all $j \neq i$, so that $\langle x, y\rangle \in \bigcap_{j\neq i}\Gamma_j = \Gamma^i$, while $\pi_i\varphi(y) = \pi_i\varphi(z)$ so that $\langle y, z\rangle \in \Gamma_i$. Therefore, $\langle x, z\rangle \in \Gamma^i \circ \Gamma_i$. Since $\langle x, z\rangle$ is an arbitrary element of U_A, it follows that $\Gamma^i \circ \Gamma_i = U_A$.

The main purpose of this section is to prove a converse of Proposition 1.4.

1.5 PROPOSITION. Let $\mathbf{A}$ be a partial algebra. Suppose that $\{\Gamma_i \mid i \in I\}$ is a subset of $\Theta(\mathbf{A})$. Define

$$\Gamma^i = \bigcap_{j\neq i}\Gamma_j.$$

Assume that

(i) $\bigcap_{i\in I}\Gamma_i = I_A$, and
(ii) $\Gamma^i \circ \Gamma_i = U_A$ for all $i \in I$.

Then there is a decomposition of $\mathbf{A}$ as an interdirect product of $\{\mathbf{A}/\Gamma_i \mid i \in I\}$.

Proof. By 2.1.4, there is a subdirect decomposition $\langle\varphi; \Pi_{i\in I}\mathbf{A}/\Gamma_i\rangle$ of $\mathbf{A}$ such that $\bar{\Gamma}_i = \pi_i \circ \varphi$ for all $i \in I$. Our objective is to prove that this is an interdirect decomposition. Let $x \in A$, and suppose that a is an element of $\mathbf{X}_{i\in I}A/\Gamma_i$ such that $|\{i \in I \mid \pi_i(a) \neq \pi_i(\varphi(x))\}| = n$. We prove by induction on n that there is an element $y \in A$ such that $a = \varphi(y)$. If $n = 0$, then $a = \varphi(x)$, so that there is nothing to prove in this case. Suppose that the assertion is true for $n - 1$. Let $\{i \in I \mid \pi_i(a) \neq \pi_i(\varphi(x))\} = \{i_0, i_1, \cdots, i_{n-1}\}$. Define b in $\mathbf{X}_{i\in I}A/\Gamma_i$ by $\pi_j(b) = \pi_j(a)$ for $j \neq i_0$, and $\pi_{i_0}(b) = \pi_{i_0}(\varphi(x))$. Then $|\{i \in I \mid \pi_i(b) \neq \pi_i(\varphi(x))\}| = n - 1$. By the induction hypothesis, there is a $z \in A$ such that $b = \varphi(z)$. Choose $w \in A$ so that $\bar{\Gamma}_{i_0}(w) = \pi_{i_0}(a)$. Then $\langle z, w\rangle \in U_A$ so that by (ii), there is an element $y \in A$ such that

$$\langle z, y\rangle \in \Gamma^{i_0} = \bigcap_{j\neq i_0}\Gamma_j,\ \langle y, w\rangle \in \Gamma_{i_0}.$$

Hence, $\pi_j(\varphi(y)) = \bar{\Gamma}_j(y) = \bar{\Gamma}_j(x) = \pi_j(\varphi(z)) = \pi_j(b) = \pi_j(a)$ for $j \neq i_0$, and $\pi_{i_0}(\varphi(y)) = \bar{\Gamma}_{i_0}(y) = \bar{\Gamma}_{i_0}(w) = \pi_{i_0}(a)$. Hence $\pi_i(\varphi(y)) = \pi_i(a)$ for all i. Thus, the induction is complete, and the proposition follows.

1.6 COROLLARY. (*Chinese remainder theorem*) Let $\mathbf{A}$ be an abstract algebra. Let $\{\Gamma_0, \Gamma_1, \cdots, \Gamma_{n-1}\} \subseteq \Theta(\mathbf{A})$ be such that

(i) $\bigcap_{i<n}\Gamma_i = I_A$.
(ii) $(\bigcap_{j\neq i}\Gamma_j) \circ \Gamma_i = U_A$ for all $i < n$.

Then $\mathbf{A}$ is isomorphic to the direct product $\Pi_{i<n}(\mathbf{A}/\Gamma_i)$.

1.7 COROLLARY. Let $\mathbf{A}$ be an abstract algebra. Then $\mathbf{A}$ is indecomposable if and only if there is no pair Γ, Δ in $\Theta(\mathbf{A})$ with $\Gamma \supset I_A$, $\Delta \supset I_A$, $\Gamma \cap \Delta = I_A$, and $\Gamma \circ \Delta = U_A$.

Proof. If such a pair of congruence exists, then by 1.6, $\mathbf{A}$ admits a proper decomposition as a direct product of $\{\mathbf{A}/\Gamma, \mathbf{A}/\Delta\}$. Conversely, suppose that $\mathbf{A}$ admits a proper decomposition φ as a direct product of $\{\mathbf{A}_i \mid i \in I\}$. Let Γ_i be the kernel of $\pi_i \circ \varphi$. By 2.1.3, $\bigcap_{i\in I}\Gamma_i = I_A$, and $\Gamma_i \supset I_A$ for $i \in I$.

Select $i \in I$ such that $\Gamma_i \neq U_A$. This is possible because $\bigcap_{i \in I} \Gamma_i = I_A$. Define $\Delta = \Gamma_i$, $\Gamma = \bigcap_{j \neq i} \Gamma_j$. Then $\Delta \supset I_A$, $\Gamma \cap \Delta = I_A$. Moreover, since every direct product is an interdirect product, it follows from 1.4 that $\Gamma \circ \Delta = U_A$. Finally, $\Gamma \supset I_A$, since $\Delta \subset U_A$.

The results in 1.6 and 1.7 do not entirely fulfill our promise to relate interdirect decompositions to the lattice structure of $\Theta(\mathbf{A})$. The composition operation for binary relations cannot in general be expressed in terms of the lattice operations of $\Theta(\mathbf{A})$. Indeed, $\Theta(\mathbf{A})$ may not even be closed under composition. That is, for Γ and Δ in $\Theta(\mathbf{A})$, the composition $\Gamma \circ \Delta$ may not be a congruence relation, or even an equivalence relation. It is not difficult to show that if $\mathbf{A}$ is an algebra, then $\Gamma \circ \Delta$ is a congruence relation if and only if $\Gamma \circ \Delta = \Delta \circ \Gamma$. To prove this, it is convenient to introduce the inverse of a binary relation P by

$$P^{-1} = \{\langle x, y \rangle \mid \langle y, x \rangle \in P\}.$$

A binary relation P on a set X is clearly an equivalence relation if and only if

$$P \supseteq I_X, \quad P^{-1} = P, \quad \text{and} \quad P \circ P = P.$$

1.8 LEMMA. Let Γ and Δ be congruence relations on an algebra $\mathbf{A}$. Then $\Gamma \circ \Delta \in \Theta(\mathbf{A})$ if and only if $\Gamma \circ \Delta = \Delta \circ \Gamma$. In this case, $\Gamma \circ \Delta$ the least upper bound of $\{\Gamma, \Delta\}$ in $\Theta(\mathbf{A})$.

Proof. Let $\mathbf{A} = \langle A; F_\xi \rangle_{\xi < p}$ be of type $\tau \in \mathrm{Ord}^p$. Suppose that $\Gamma \circ \Delta \in \Theta(\mathbf{A})$. Then in particular $\Gamma \circ \Delta$ is an equivalence relation. Therefore, $\Gamma \circ \Delta = (\Gamma \circ \Delta)^{-1} = \Delta^{-1} \circ \Gamma^{-1} = \Delta \circ \Gamma$, using the easily verified identity $(\Gamma \circ \Delta)^{-1} = \Delta^{-1} \circ \Gamma^{-1}$. Conversely, suppose that $\Gamma \circ \Delta = \Delta \circ \Gamma$. Then $\Gamma \circ \Delta \supseteq I_A \circ I_A = I_A$, $(\Gamma \circ \Delta)^{-1} = \Delta^{-1} \circ \Gamma^{-1} = \Delta \circ \Gamma = \Gamma \circ \Delta$, and $(\Gamma \circ \Delta) \circ (\Gamma \circ \Delta) = (\Gamma \circ \Delta) \circ (\Delta \circ \Gamma) = \Gamma \circ ((\Delta \circ \Delta) \circ \Gamma) = \Gamma \circ (\Delta \circ \Gamma)$ $= \Gamma \circ (\Gamma \circ \Delta) = (\Gamma \circ \Gamma) \circ \Delta = \Gamma \circ \Delta$. Therefore, $\Gamma \circ \Delta$ is an equivalence relation. Suppose that $\mathbf{a}$, $\mathbf{b}$ in $A^{\tau(\xi)}$ are such that $\langle \mathbf{a}(\eta), \mathbf{b}(\eta) \rangle \in \Gamma \circ \Delta$ for all $\eta < \tau(\xi)$. Then $\mathbf{c} \in A^{\tau(\xi)}$ exists satisfying $\langle \mathbf{a}(\eta), \mathbf{c}(\eta) \rangle \in \Gamma$, $\langle \mathbf{c}(\eta), \mathbf{b}(\eta) \rangle \in \Delta$ for all $\eta < \tau(\xi)$. Consequently, $\langle F_\xi(\mathbf{a}), F_\xi(\mathbf{c}) \rangle \in \Gamma$ and $\langle F_\xi(\mathbf{c}), F_\xi(\mathbf{b}) \rangle \in \Delta$. Therefore, $\langle F_\xi(\mathbf{a}), F_\xi(\mathbf{b}) \rangle \in \Gamma \circ \Delta$. It is this step of the proof that requires the assumption that $\mathbf{A}$ is an algebra. Otherwise, $F_\xi(\mathbf{c})$ might not be defined. The argument up to this point shows that $\Gamma \circ \Delta$ is a congruence relation on $\mathbf{A}$. To prove that $\Gamma \circ \Delta = \Gamma \vee \Delta$ in $\Theta(\mathbf{A})$, note that $\Gamma \circ \Delta \supseteq \Gamma \circ I_A = \Gamma$ and $\Gamma \circ \Delta \supseteq I_A \circ \Delta = \Delta$. Hence, $\Gamma \circ \Delta$ is an upper bound of $\{\Gamma, \Delta\}$. If Γ_1 is a congruence relation on $\mathbf{A}$ such that $\Gamma_1 \supseteq \Gamma$ and $\Gamma_1 \supseteq \Delta$, then Γ_1 $= \Gamma_1 \circ \Gamma_1 \supseteq \Gamma \circ \Delta$. Thus, $\Gamma \circ \Delta = \Gamma \vee \Delta$.

It follows from 1.8 that if all of the congruence relations of an algebra $\mathbf{A}$ commute, then the investigation of interdirect decompositions of $\mathbf{A}$ can be carried out as an arithmetical study of the lattice $\Theta(\mathbf{A})$. In fact, when all congruences commute, $\Gamma \circ \Delta = \Gamma \vee \Delta$ for each Γ and Δ in $\Theta(\mathbf{A})$. Therefore, by Lemma 2.2.3 and Propositions 1.4 and 1.5, the interdirect decompositions of $\mathbf{A}$ are in one-to-one correspondence with the subsets $\{\Gamma_i \mid i \in I\}$ of $\Theta(\mathbf{A})$

which satisfy

(i) $\bigwedge_{i \in I} \Gamma_i = I_A$, and
(ii) $(\bigwedge_{j \neq i} \Gamma_j) \vee \Gamma_i = U_A$ for all $i \in I$.

Since these conditions involve only the lattice operations of $\Theta(\mathbf{A})$, it follows that the interdirect decompositions of $\mathbf{A}$ are completely determined by the lattice structure of $\Theta(\mathbf{A})$.

If $\Gamma \circ \Delta = \Delta \circ \Gamma$ for all Γ and Δ in $\Theta(\mathbf{A})$, then $\mathbf{A}$ is said to be an *algebra with commuting congruences*. This seems like a rather unnatural condition. However, many kinds of algebras have commuting congruences.

1.9 EXAMPLES. (a) Let $\mathbf{G} = \langle G; \times, 1, ^{-1}, F_\xi \rangle_{\xi < \rho}$ be a group with unary operators satisfying $F_\xi(x \times y) = F_\xi(x) \times F_\xi(y)$ for all x and y in G. It is well known that every congruence relation on $\mathbf{G}$ is of the form

$$\Gamma_N = \{\langle x, y \rangle \mid x \times y^{-1} \in N\},$$

where $N \subseteq G$ is a subgroup of $\mathbf{G}$ such that

$$x \times N \times x^{-1} \subseteq N \quad \text{for} \quad x \in G, \qquad F_\xi(N) \subseteq N \quad \text{for} \quad \xi < \rho.$$

Let M and N be two such subgroups. Suppose that $\langle x, y \rangle \in \Gamma_N \circ \Gamma_M$. Then there exists $z \in G$ such that $x \times z^{-1} \in N, z \times y^{-1} \in M$. Let $w = x \times z^{-1} \times y$. Then $x \times w^{-1} = (x \times y^{-1}) \times (z \times y^{-1}) \times (x \times y^{-1})^{-1} \in M$ and $w \times y^{-1} = x \times z^{-1} \in N$. Hence, $\langle x, y \rangle \in \Gamma_M \circ \Gamma_N$. Therefore, $\Gamma_N \circ \Gamma_M \subseteq \Gamma_M \circ \Gamma_N$. By symmetry, $\Gamma_M \circ \Gamma_N = \Gamma_N \circ \Gamma_M$. Thus, $\mathbf{G}$ has commuting congruences.

(b) (Dilworth [14]). A lattice $\mathbf{L}$ is called *relatively complemented* if for any elements $u \leqq z \leqq v$ in L, there exists $w \in L$ such that

$$z \wedge w = u, \qquad z \vee w = v.$$

We prove that every relatively complemented lattice has commuting congruences. Suppose that $\langle x, y \rangle \in \Gamma \circ \Delta$, where Γ and Δ are congruence relations on the relatively complemented lattice $\mathbf{L}$. Then z exists in L such that

(i) $x \,\Gamma\, z$, and
(ii) $z \,\Delta\, y$.

Since $\mathbf{L}$ is relatively complemented, elements u and v exist in L such that

(iii) $(x \vee z) \vee u = x \vee y \vee z$.
(iv) $(x \vee z) \wedge u = x$.
(v) $(y \vee z) \vee v = x \vee y \vee z$.
(vi) $(y \vee z) \wedge v = y$.

Define $w = u \wedge v$. By (i) and (v)

$$(y \vee z) \,\Gamma\, (x \vee y \vee z) = (y \vee z) \vee v.$$

Thus, by (vi)

(vii) $y = (y \vee z) \wedge v \,\Gamma\, ((y \vee z) \vee v) \wedge v = v$.

Similarly,

(viii) $x \mathrel{\Delta} u$.

By (i) and (iv),

$$x \vee z \mathrel{\Gamma} x = (x \vee z) \wedge u.$$

Therefore, by (iii)

$$x \vee y \vee z = (x \vee z) \vee u \mathrel{\Gamma} ((x \vee z) \wedge u) \vee u = u.$$

Since $x \vee y \vee z \geqq u, v$ by (iii) and (v), it follows that

$$x \vee y \vee z \geqq u \vee v \geqq u.$$

Consequently,

$$u \vee v = (x \vee y \vee z) \wedge (u \vee v) \mathrel{\Gamma} u \wedge (u \vee v) = u.$$

Hence,

$$v = (u \vee v) \wedge v \mathrel{\Gamma} u \wedge v = w.$$

Similarly, $u \mathrel{\Delta} w$. Thus, by (vii) and (viii)

$$y \mathrel{\Gamma} w, \qquad x \mathrel{\Delta} w,$$

so that

$$\langle x, y \rangle \in \Delta \circ \Gamma.$$

We have shown $\Gamma \circ \Delta \subseteq \Delta \circ \Gamma$. By symmetry, $\Delta \circ \Gamma \subseteq \Gamma \circ \Delta$. Therefore, every relatively complemented lattice has commuting congruence relations.

The assumption that an algebra **A** has commuting congruence relations imposes an important restriction on $\Theta(\mathbf{A})$.

1.10 PROPOSITION. Let **A** be an algebra with commuting congruence relations. Suppose that Γ, Δ, E are elements of $\Theta(\mathbf{A})$ such that $\Gamma \supseteq E$. Then

$$\Gamma \wedge (\Delta \vee E) = (\Gamma \wedge \Delta) \vee E.$$

Proof. Clearly, $\Gamma \wedge (\Delta \vee E) \supseteq \Gamma \wedge \Delta$. Also, since $\Gamma \supseteq E$, it also follows that $\Gamma \wedge (\Delta \vee E) \supseteq E$. Thus, $\Gamma \wedge (\Delta \vee E) \supseteq (\Gamma \wedge \Delta) \vee E$. Let $\langle x, y \rangle \in \Gamma \wedge (\Delta \vee E) = \Gamma \cap (\Delta \circ E)$. Then $\langle x, y \rangle \in \Gamma$ and there exists z such that $\langle x, z \rangle \in \Delta$, $\langle z, y \rangle \in E \subseteq \Gamma$. Hence, $\langle x, z \rangle \in \Gamma$ (since $\langle x, y \rangle \in \Gamma$, $\langle y, z \rangle \in \Gamma$, and Γ is an equivalence relation). Therefore, $\langle x, z \rangle \in \Gamma \cap \Delta$. Thus, $\langle x, y \rangle \in (\Gamma \cap \Delta) \circ E = (\Gamma \wedge \Delta) \vee E$. This proves that $\Gamma \wedge (\Delta \vee E) \subseteq (\Gamma \wedge \Delta) \vee E$, which is the required inclusion.

2. Independence in Modular Lattices

The purpose of the remaining sections of this chapter is to prove a fundamental uniqueness theorem for direct decompositions of abstract algebras. The discussion in the preceding section shows that the finite direct decompositions of an abstract algebra **A** are determined by the lattice structure of

$\Theta(\mathbf{A})$, provided **A** has commuting congruences. The uniqueness theorem is obtained from a result of Ore on the arithmetic of certain lattices. This section and the following one provide the background of lattice theory needed for the proof of Ore's theorem which is given in section 4.

2.1 DEFINITION. A lattice **L** is *modular* if it satisfies the *modular law*: $x \geqq z$ implies $x \wedge (y \vee z) = (x \wedge y) \vee z$ for all x, y, and z in L.

By 1.10 if **A** is an algebra with commuting congruence relations, then $\Theta(\mathbf{A})$ is a modular lattice. This fact makes the study of modular lattices important for the theory of abstract algebras. In this section and the following one, some of the properties of modular lattices will be presented.

If $b \leqq a$ are two elements of a lattice **L**, then a/b denotes the interval $\{x \in L \mid b \leqq x \leqq a\}$. As a subset of L, a/b inherits a partial ordering, and it is easy to see that least upper bounds and greatest lower bounds are the same for subsets of a/b as they are in **L**. In particular, a/b is a sublattice of **L**, called the *quotient sublattice* determined by a and b (or simply the quotient of a by b).

2.2 LEMMA. (*transposition principal*) Let **L** be a modular lattice, and let a and b be elements of L. Then the mappings $\lambda : x \to a \vee x$ and $\mu : y \to b \wedge y$ are inverse isomorphisms between $b/(a \wedge b)$ and $(a \vee b)/a$.

Proof. If $a \wedge b \leqq x \leqq b$, then $a = a \vee (a \wedge b) \leqq a \vee x \leqq a \vee b$, and if $a \leqq y \leqq a \vee b$, then $a \wedge b \leqq y \wedge b \leqq (a \vee b) \wedge b = b$. Thus, λ maps $b/a \wedge b$ to $a \vee b/a$, and μ maps $a \vee b/a$ to $b/a \wedge b$. Clearly, $x_1 \leqq x_2$ in $b/(a \wedge b)$ implies $\lambda(x_1) \leqq \lambda(x_2)$ and $y_1 \leqq y_2$ in $(a \vee b)/a$ implies $\mu(y_1) \leqq \mu(y_2)$. Finally, if $a \wedge b \leqq x \leqq b$, then $\mu(\lambda(x)) = b \wedge (a \vee x) = (b \wedge a) \vee x = x$ by the modular law. Similarly, $\lambda(\mu(y)) = y$ for all y such that $a \leqq y \leqq a \vee b$.

Two quotients x/y and z/w are called transposes of each other if they are related as in this lemma, that is, if either $y = x \wedge w$ and $z = x \vee w$, or $w = y \wedge z$ and $x = y \vee z$. The transposition principal asserts that transposed quotients are isomorphic.

The abstract study of direct decompositions is based on a definition, which is motivated by 1.6.

2.3 DEFINITION. Let **L** be a modular lattice with a largest element u. A set $\{x_0, x_1, \cdots, x_{n-1}\}$ of elements of L is called *meet independent* if

$$x_i \vee (\textstyle\bigwedge_{j \neq i} x_j) = u \quad \text{for all} \quad i < n.$$

If **L** has a smallest element o, then a set $\{y_0, y_1, \cdots, y_{n-1}\}$ in L is called *join independent* if

$$y_i \wedge (\textstyle\bigvee_{j \neq i} y_j) = 0 \quad \text{for all} \quad i < n.$$

2.4 LEMMA. Let **L** be a modular lattice with a least element o. Suppose that $\{y_0, y_1, \cdots, y_{n-1}\}$ is a join independent subset in L. Then for any two subsets S and T of n,

$$(\bigvee_{i\in S}y_i) \wedge (\bigvee_{i\in T}y_i) = \bigvee_{i\in S\cap T}y_i.$$

If $\mathbf{L}$ is a modular lattice with a greatest element u, and $\{x_0, x_1, \cdots, x_{n-1}\}$ is a meet independent subset of L, then for any two subsets S and T of n,

$$(\bigwedge_{i\in S}x_i) \vee (\bigwedge_{i\in T}x_i) = \bigwedge_{i\in S\cap T}x_i.$$

(Note that by definition, the least upper bound of the empty set is o, and the greatest lower bound of the empty set is u.)

Proof. We will prove the first part of the lemma. The second part can be established by the same argument, with joins and meets interchanged.† Suppose first that $S \cap T = \varnothing$. What is to be shown is $(\bigvee_{i\in S}y_i) \wedge (\bigvee_{i\in T}y_k) = o$. If $S = \varnothing$, then $\bigvee_{i\in S}y_i = o$, which gives the desired result. It can therefore be assumed that $S \neq \varnothing$, and that the identity holds for sets which are smaller than S. Let $j \in S$, and put $S' = S - \{j\}$, $T' = T \cup \{j\}$. Then by the modular law and the induction hypothesis,

$$\begin{aligned}(\bigvee_{i\in S}y_i) \wedge (\bigvee_{i\in T}y_i) &\leqq (\bigvee_{i\in S}y_i) \wedge (\bigvee_{i\in T'}y_i)\\ &= (y_j \vee \bigvee_{i\in S'}y_i) \wedge (\bigvee_{i\in T'}y_i) = y_j \vee ((\bigvee_{i\in S'}y_i) \wedge (\bigvee_{i\in T'}y_i))\\ &= y_j \vee o = y_j.\end{aligned}$$

Hence,

$$(\bigvee_{i\in S}y_i) \wedge (\bigvee_{i\in T}y_i) \leqq y_j \wedge (\bigvee_{i\neq j}y_i) = o.$$

The general case follows easily from the special case by the modular law:

$$\begin{aligned}&(\bigvee_{i\in S}y_i) \wedge (\bigvee_{i\in T}y_i)\\ &\quad= ((\bigvee_{i\in S\cap T}y_i) \vee (\bigvee_{i\in S-S\cap T}y_i)) \wedge (\bigvee_{i\in T}y_i)\\ &\quad= (\bigvee_{i\in S\cap T}y_i) \vee ((\bigvee_{i\in S-S\cap T}y_i) \wedge (\bigvee_{i\in T}y_i))\\ &\quad= (\bigvee_{i\in S\cap T}y_i) \vee o = \bigvee_{i\in S\cap T}y_i.\end{aligned}$$

2.5 COROLLARY. Let $\mathbf{L}$ be a modular lattice with a greatest element u and a least element o. Suppose that $\{y_0, y_1, \cdots, y_{n-1}\}$ is a join independent set such that $\bigvee_{i<n}y_i = u$. For $i < n$, define

$$\bar{y}_i = \bigvee_{j\neq i}y_j.$$

Then $\{\bar{y}_0, \bar{y}_1, \cdots, \bar{y}_{n-1}\}$ is meet independent and $\bigwedge_{i<n}\bar{y}_i = o$. Suppose that

† The second part of the lemma can also be obtained from the first part by applying the "principle of duality." Corresponding to each definition or statement expressed in terms of the order relation $\leqq$ of a partially ordered set, there is a dual definition or statement obtained by replacing each occurrence of $\leqq$ by $\geqq$ and each concept defined in terms of $\leqq$ by its dual. For example, the dual of "join" is "meet" and vice versa, and the dual of "meet independent" is "join independent" and vice versa. The principle of duality asserts that if a statement holds for all partially ordered sets, then so does its dual. The principle is easily proved by observing that if $\mathbf{P} = \langle P; \leqq\rangle$ is a partially ordered set, and if $\mathbf{P}' = \langle P; \geqq\rangle$, then any statement about $\mathbf{P}$ is equivalent to the dual statement about $\mathbf{P}'$.

$\{x_0, x_1, \cdots, x_{n-1}\} \subseteq L$ is a meet independent set such that $\bigwedge_{i<n} x_i = o$. For $i < n$, define

$$\bar{x}_i = \bigwedge_{j \neq i} x_j.$$

Then $\{\bar{x}_0, \bar{x}_1, \cdots, \bar{x}_{n-1}\}$ is join independent and $\bigvee_{i<n} \bar{x}_i = u$.

Proof. Evidently, $y_i \leqq \bar{y}_j$ if $j \neq i$. Therefore, $\bigwedge_{j \neq i} \bar{y}_j \geqq y_i$. Consequently, $(\bigwedge_{j \neq i} \bar{y}_j) \vee \bar{y}_i \geqq y_i \vee \bar{y}_i = \bigvee_{j<n} y_j = u$. Thus $\{\bar{y}_0, \bar{y}_1, \cdots, \bar{y}_{n-1}\}$ is meet independent. By 2.4, $\bigwedge_{i<n} \bar{y}_i = \bigwedge_{i<n} (\bigvee_{j \neq i} y_j) = \bigvee_{i \in \phi} y_i = o$. The second statement of the theorem can be proved by a similar argument, or else it can be inferred from the principle of duality.

This corollary shows that the concepts of join and meet independence can be used interchangably. It is slightly more convenient to develop the theory of join independence rather than meet independence, even though this theory does not directly yield results on the structure of algebras. However, the following lemma can be used to carry over results on join independence to give information on direct decompositions.

2.6 LEMMA. Let $\mathbf{A} = \langle A; F_\xi \rangle_{\xi<\rho}$ be an algebra with commuting congruence relations. Suppose that $\{\Delta_0, \Delta_1, \cdots, \Delta_{n-1}\}$ is a join independent subset of $\Theta(\mathbf{A})$ such that $\bigvee_{i<n} \Delta_i = U_A$. Define $\Gamma_i = \bigvee_{j \neq i} \Delta_j$. Then $\mathbf{A}$ is isomorphic to $\Pi_{i<n} \mathbf{A}/\Gamma_i$. Moreover, $\Theta(\mathbf{A}/\Gamma_i)$ is lattice isomorphic to Δ_i/I_A.

Everything required for the proof of this lemma has been done. The appropriate references are 2.3.3, 1.6, 1.10, 2.2, and 2.5 in suitable order.

Henceforth, the term "independent set" will designate a join independent set in a modular lattice with a least element o.

2.7 LEMMA. Let $\{x_0, x_1, \cdots, x_{n-1}\}$ be independent. Then any non-empty subset of $\{x_0, x_1, \cdots, x_{n-1}\}$ is independent. Moreover, if $\{y, z\}$ is independent, and $y \vee z = x_0$, then $\{y, z, x_1, \ldots, x_{n-1}\}$ is an independent set.

Proof. The first statement is evident from the definition of join independence. By symmetry and the known independence of $\{x_0, x_1, \cdots, x_{n-1}\}$. it is sufficient to show

$$y \wedge (z \vee \bigvee_{0<i<n} x_i) = o.$$

Since $x_0 \wedge \bigvee_{0<i<n} x_i = o$, $y \wedge z = o$, and $y \vee z = x_0$, the modular law yields $y \wedge (z \vee \bigvee_{0<i<n} x_i) = y \wedge [x_0 \wedge (z \vee \bigvee_{0<i<n} x_i)] = y \wedge [z \vee (x_0 \wedge \bigvee_{0<i<n} x_i)] = y \wedge z = 0$.

This lemma justifies the introduction of some convenient notation.

2.8 DEFINITION. Let $\mathbf{L}$ be a modular lattice with least element o. If $a \in L$ and $a = x_0 \vee x_1 \vee \cdots \vee x_{n-1}$, where $\{x_0, x_1, \cdots, x_{n-1}\}$ is an independent set, then a is called the direct join of $\{x_0, x_1, \cdots, x_{n-1}\}$, and it is customary to write

$$a = x_0 \otimes x_1 \otimes \cdots \otimes x_{n-1}.$$

If $a = x \otimes y$, then x (and also y) is called a *direct factor* of a. An element

a in L is called *indecomposable* (in **L**) if $a \neq o$ and the only direct factors of a are o and a.

By 2.7, $\otimes$ can be considered as an associative, binary partial operation on L. Clearly, $\otimes$ is also commutative. Note that an expression such as $x_0 \otimes x_1 \otimes \cdots \otimes x_{n-1}$ not only stands for the element $x_0 \vee x_1 \vee \cdots \vee x_{n-1}$, but it also means that $\{x_0, x_1, \cdots, x_{n-1}\}$ is independent.

We will say that a modular lattice **L** with a least element o satisfies the *restricted chain condition* if L contains no infinite sequence

$$x_0 > x_1 > x_2 > x_3 > \cdots,$$

where x_{n+1} is a direct factor of x_n for all $n < \omega$. This condition is clearly satisfied if **L** satisfies either the descending chain condition (L contains no infinite sequence $y_0 > y_1 > y_2 > \cdots$), or the ascending chain condition (L contains no infinite sequence $z_0 < z_1 < z_2 < \cdots$). Moreover, there are lattices which satisfy neither of these conditions, but do satisfy the restricted chain condition (see Problem 2).

2.9 PROPOSITION. Let **L** be a modular lattice with zero element o. Suppose that **L** satisfies the restricted chain condition. Then every nonzero element of L is a direct join of indecomposable elements.

Proof. (i) Let $a > o$ in **L**. Then it is possible to write $a = y \otimes x$, where x is indecomposable in **L**. In fact, suppose otherwise. Then a is not indecomposable. Hence, $a = y_0 \otimes x_1$, where $a > x_1 > o$. By assumption, x_1 is not indecomposable, so that $x_1 = y_1 \otimes x_2$, where $x_1 > x_2 > 0$. By 2.7, x_2 is a direct factor of a, so the assumption that a has no indecomposable direct factors implies that x_2 is decomposable. This reasoning can be repeated indefinitely to obtain a sequence $a = x_0 > x_1 > x_2 > \cdots$ of the kind that is forbidden by the restricted chain condition.

(ii) Let $a > o$ in **L**. By (i), there is an indecomposable element x_0 in L such that $a = x_0 \otimes y_0$ for some y_0. If $y_0 = o$, then $a = x_0$ is indecomposable. If $y_0 > o$, then by (i), $y_0 = x_1 \otimes y_1$, where x_1 is indecomposable. If $y_1 = o$, $a = x_0 \otimes x_1$. If $y_1 > o$, then $y_1 = x_2 \otimes y_2$, where x_2 is indecomposable. This process can be continued as long as $y_n > o$. By the restricted chain condition, it must terminate, since otherwise an infinite sequence $y_0 > y_1 > y_2 > \cdots$ will be obtained in which each y_{n+1} is a direct factor of y_n.

2.10 THEOREM. Let $\mathbf{A} = \langle A; F_\xi \rangle_{\xi < \rho}$ be an algebra with commuting congruence relations, such that $\Theta(\mathbf{A})$ satisfies the restricted chain condition. Then **A** is isomorphic to a finite direct product of indecomposable algebras.

Proof. By 2.9, it is possible to write $U_A = \Delta_0 \otimes \Delta_1 \otimes \cdots \otimes \Delta_{n-1}$, where each Δ_i is indecomposable in $\Theta(\mathbf{A})$. Corresponding to this direct join, there is a decomposition of **A** as a direct product $\Pi_{i<n}\mathbf{A}_i$, where $\Theta(\mathbf{A}_i)$ is isomorphic to Δ_i/I_A. It follows from 1.7 and the fact that Δ_i is indecomposable in $\Theta(\mathbf{A})$ that the algebra $\mathbf{A}_i$ is indecomposable.

3. Finite Dimensional Modular Lattices

The principal tool in the proof of the uniqueness theorem for direct decompositions is the dimension function on the lattice of congruence relations. Such functions are the lattice theoretical generalization of the dimension function defined on the subspaces of a finite dimensional vector space.

Dimension functions do not exist for all modular lattices. One can of course simply assume that the algebra **A** under consideration is such that $\Theta(\mathbf{A})$ has a dimension function defined on it. Alternatively, more natural restrictions can be imposed on $\Theta(\mathbf{A})$ that imply the existence of a dimension function. The theory developed in this way has a number of useful applications in algebra, so that it seems worthwhile to present this development. The reader who wishes to do so can omit this section.

3.1 DEFINITION. Let **L** be a lattice with a least element o. A *dimension function* on **L** is a mapping d of L into the integers satisfying

(i) $d(x) + d(y) = d(x \wedge y) + d(x \vee y)$.
(ii) $x < y$ implies $d(x) < d(y)$.
(iii) $d(o) = 0$.

A lattice **L** is called *finite dimensional* if **L** has a least element and a greatest element and there is a dimension function defined on **L**.

The existence of a dimension function on a lattice **L** is very restrictive. For example, **L** must be modular. In fact, if $x \geqq z$, then $x \geqq (x \wedge y) \vee z$ and $y \vee z \geqq (x \wedge y) \vee z$. Therefore $x \wedge (y \vee z) \geqq (x \wedge y) \vee z$. By repeated use of 3.1 (i), we obtain $d(x \wedge (y \vee z)) = d(x) + d(y \vee z) - d(x \vee y \vee z) = d(x) + d(y \vee z) - d(x \vee y)$, and $d((x \wedge y) \vee z) = d(x \wedge y) + d(z) - d(x \wedge y \wedge z) = d(x) + d(y) - d(x \vee y) + d(z) - d(y \wedge z) = d(x) + (d(y) + d(z) - d(y \wedge z)) - d(x \vee y) = d(x) + d(y \vee z) - d(x \vee y) = d(x \wedge (y \vee z))$. Therefore, by 3.1 (ii), $x \wedge (y \vee z) = (x \wedge y) \vee z$. This proves that **L** is modular. In addition to being modular, **L** must satisfy the *descending chain condition*: if $x_0 \geqq x_1 \geqq x_2 \geqq \cdots$ in **L**, then there is an n such that $x_n = x_{n+1} = x_{n+2} = \cdots$. This fact is obvious from 3.1 (ii), since an infinite sequence $x_0 > x_1 > x_2 > \cdots$ would yield an infinite decreasing sequence of non-negative integers: $d(x_0) > d(x_1) > d(x_2) > \cdots$. If **L** has a largest element u, then $d(x) \leqq d(u)$ for all $x \in L$, and by the same argument **L** satisfies the *ascending chain condition*: if $x_0 \leqq x_1 \leqq x_2 \leqq \cdots$ in **L**, then there is an n such that $x_n = x_{n+1} = x_{n+2} = \cdots$.

The discussion in the preceding paragraph proves half of the following theorem which characterizes finite dimensional lattices.

3.2 THEOREM. A lattice **L** is finite dimensional if and only if **L** is modular and satisfies both the ascending chain condition and the descending chain condition.

The proof that every modular lattice that satisfies both chain conditions is finite dimensional uses the lattice theoretical analogue of the Jordan-Hölder theorem. This result is obtained from the "Zassenhaus Lemma."

3.3 LEMMA. (Zassenhaus). Let $x_1 \leqq x_2$ and $y_1 \leqq y_2$ in the modular lattice **L**. Then the quotients $(x_1 \vee (x_2 \wedge y_2))/(x_1 \vee (x_2 \wedge y_1))$ and $(y_1 \vee (y_2 \wedge x_2))/(y_1 \vee (y_2 \wedge x_1))$ are transposes of $(x_2 \wedge y_2)/((x_1 \wedge y_2) \vee (y_1 \wedge x_2))$ and, therefore, are isomorphic.

Proof. Let $a = x_2 \wedge y_2$, $b = x_1 \vee (x_2 \wedge y_1)$. Then

$$\begin{aligned} a \wedge b &= (x_2 \wedge y_2) \wedge (x_1 \vee (x_2 \wedge y_1)) \\ &= (x_2 \wedge y_2 \wedge x_1) \vee (x_2 \wedge y_1) \\ &= (y_2 \wedge x_1) \vee (x_2 \wedge y_1). \end{aligned}$$

Also, $a \vee b = (x_2 \wedge y_2) \vee x_1 \vee (x_2 \wedge y_1) = (x_2 \wedge y_2) \vee x_1$. Thus, $(x_1 \vee (x_2 \wedge y_2))/(x_1 \vee (x_2 \wedge y_1))$ is a transpose of $(x_2 \wedge y_2)/((y_2 \wedge x_1) \vee (x_2 \wedge y_1))$. Similarly $(y_1 \vee (y_2 \wedge x_2)/(y_1 \vee (y_2 \wedge x_1))$ is a transpose of $(x_2 \wedge y_2)/((x_2 \wedge y_1) \vee (y_2 \wedge x_1))$.

A nonempty subset C of a partially ordered set is called a *chain* if for each pair of elements x and y in C, either $x \leqq y$ or $y \leqq x$. If C contains a least element a and a greatest element b, then a and b are respectively called the lower and upper *endpoints* of C. If C and C' are two chains in the same partially ordered set such that $C \subseteq C'$, then C' is called a *refinement* of C.

3.4 LEMMA. (*Jordan-Hölder-Schreier*) Let **L** be a modular lattice. Suppose that a and b are elements of L and $a \leqq b$. Assume that $C: a = x_0 < x_1 < \cdots < x_{n-1} < x_n = b$, and $D: a = y_0 < y_1 < \cdots < y_{m-1} < y_m = b$ are chains in **L**. Let $C': a = x_{00} \leqq x_{01} \leqq \cdots \leqq x_{0m} \leqq x_{10} \leqq x_{11} \leqq \cdots, \cdots, \cdots \leqq x_{n-1\,0} \leqq \cdots \leqq x_{n-1\,m} = b$ and $D': a = y_{00} \leqq y_{01} \leqq \cdots \leqq y_{0m} \leqq y_{10} \leqq y_{11} \cdots, \cdots, \cdots \leqq y_{m-1\,0} \leqq \cdots \leqq y_{m-1\,n} = b$ where

$$\begin{aligned} x_{ij} &= x_i \vee (x_{i+1} \wedge y_j), \\ y_{ij} &= y_i \vee (y_{i+1} \wedge x_j). \end{aligned}$$

Then C' and D' are chains refining C and D respectively, such that $x_{i\,j+1}/x_{ij}$ is isomorphic to $y_{j\,i+1}/y_{ji}$ for $i < n, j < m$, and $x_{im} = x_{i+1} = x_{i+1\,0}$, $y_{in} = y_{i+1} = y_{i+1\,0}$.

This result is an immediate consequence of the Zassenhaus lemma.

3.5 PROPOSITION. If **L** is a modular lattice that satisfies the ascending chain condition and the descending chain condition, then for any $a \leqq b$ in **L** there is a finite maximal chain having endpoints a and b. Any two maximal chains having the same endpoints contain the same number of distinct elements.

Proof. It is easily seen that the set of all chains having a and b as endpoints satisfies the conditions of Zorn's lemma. Hence, there is a maximal chain C with these endpoints. Suppose that C is infinite. If S is a nonempty subset of C, then S has a smallest element (in the ordering of **L**). Indeed, otherwise it is possible to choose $x_0, x_1, x_2, \cdots$ in S inductively, such that $x_0 > x_1 > x_2 > \cdots$. However, such a sequence violates the descending chain condition in **L**. Let y_0 be the smallest element of C. Let y_1 be the smallest element in $C - \{y_0\}$. Note that $C - \{y_0\} \neq \varnothing$ since C is supposed to be infinite. Then $y_1 > y_0$. Let y_2 be the smallest element in $C - \{y_0, y_1\}$. Then $y_2 > y_1$. Continuing this process indefinitely, we obtain a sequence $y_0 < y_1 < y_2 < \cdots$ that violates the ascending chain condition. Hence, C is finite. If C and D are maximal chains with the same endpoints, then by 3.4, C and D have refinements

$$C' = \{z_0, z_1, \cdots, z_r\} \supseteq C,$$
$$D' = \{w_0, w_1, \cdots, w_r\} \supseteq D,$$

with the same endpoints as C and D, such that for some permutation κ of $r + 1$, each z_{i+1}/z_i is isomorphic to $w_{\kappa(i)+1}/w_{\kappa(i)}$. In particular $z_{i+1} = z_i$ if and only if $w_{\kappa(i)+1} = w_{\kappa(i)}$. Thus $|C'| = |D'|$. By maximality, $C = C'$ and $D = D'$. Hence, $|C| = |D|$.

It is now possible to complete the proof of Theorem 3.2. Let **L** be a modular lattice that satisfies the ascending chain condition and the descending chain condition. Then **L** has a greatest element u, since otherwise it would be possible to choose $x_0 < x_1 < x_2 < \cdots$ in L by induction. Such a sequence violates the ascending chain condition. Similarly, **L** has a least element o. To prove that **L** is finite dimensional, it is necessary to construct a dimension function on **L**. For $a \in L$, let $d(a)$ be one less than the number of elements in a maximal chain that has o and a as its endpoints. By 3.5 this is a well-defined integer $\geqq 0$. Evidently, $d(0) = 0$. The proof that d satisfies the conditions (i) and (ii) of 3.1 uses an observation. If C is a maximal chain with endpoints o and a, and if D is a maximal chain with endpoints a and b, where $a \leqq b$, then $C \cup D$ is a maximal chain with endpoints o and b. To see this, suppose that E is a chain containing $C \cup D$. Then $E = C' \cup D'$, where $C' = \{x \in E \mid x \leqq a\}$ and $D' = \{x \in E \mid x \geqq a\}$. Clearly C' is a chain with endpoints o and a such that $C \subseteq C'$ and D' is a chain with endpoints a and b such that $D \subseteq D'$. By the maximality of C and D, it follows that $C' = C$ and $D' = D$. Hence, $E = C \cup D$. Therefore, $C \cup D$ is maximal. An important fact follows from this remark: if $a \leqq b$, then the number of elements in a maximal chain with a and b as endpoints is $d(b) - d(a) + 1$. Indeed, using the above notation, $d(b) = |C \cup D| - 1 = |C| + |D| - |C \cap D| - 1 = d(a) + 1 + |D| - |\{a\}| - 1 = d(A) + |D| - 1$, so that $|D| = d(b) - d(a) + 1$. In particular, if $a < b$, then $\{a, b\} \subseteq D$, so that $d(b) - d(a) + 1 \geqq 2$ and $d(a) < d(b)$. Suppose that x and y are arbitrary elements of L. By the transposition principle, $(x \vee y)/x$ is isomorphic to

$y/(x \wedge y)$. Therefore, a maximal chain with endpoints x and $x \vee y$ has the same number of elements as a maximal chain with endpoints $x \wedge y$ and y. Therefore,

$$d(x \vee y) - d(x) + 1 = d(y) - d(x \wedge y) + 1.$$

Transposing, we obtain

$$d(x) + d(y) = d(x \wedge y) + d(x \vee y).$$

Thus, d is a dimension function on $\mathbf{L}$ and the proof of 3.2 is finished.

4. Ore's Theorem

The lattice theoretical basis for the uniqueness of direct decompositions of an algebra is the theorem of Ore. The hypotheses under which Ore's theorem can be proved are very restrictive. However, examples show that unless strong conditions of finiteness are imposed on the lattice of congruences of an algebra $\mathbf{A}$, there may be many essentially different decompositions of $\mathbf{A}$ into a direct product of indecomposable algebras.

Throughout this section it will be assumed that $\mathbf{L}$ is a finite dimensional lattice. Thus, $\mathbf{L}$ is a modular lattice with a least element o and a greatest element u. Let d be an arbitrary dimension function on $\mathbf{L}$. This function will be fixed throughout the section.

The statement and proof of Ore's theorem is preceded by five simple lemmas concerning independence.

4.1 LEMMA. Let $a = x_0 \vee x_1 \vee \cdots \vee x_{n-1}$. Then $d(a) \leqq \Sigma_{i<n} d(x_i)$. Moreover, $a = x_0 \otimes x_1 \otimes \cdots \otimes x_{n-1}$ if and only if $d(a) = \Sigma_{i<n} d(x_i)$.

Proof. If $n = 1$, there is nothing to prove. Assume that the statement is true when n is replaced by $n - 1$. Let $b = x_1 \vee \cdots \vee x_{n-1}$. Then $a = x_0 \vee b$, so that $d(a) = d(x_0) + d(b) - d(x_0 \wedge b) \leqq d(x_0) + d(b) \leqq d(x_0) + \Sigma_{0<i<n} d(x_i) = \Sigma_{i<n} d(x_i)$. Moreover, the equality $d(a) = \Sigma_{i<n} d(x_i)$ holds if and only if $d(x_0 \wedge b) = 0$ and $d(b) = \Sigma_{0<i<n} d(x_i)$. By the Definition 3.1 of a dimension function, the condition $d(x_0 \wedge b) = 0$ is equivalent to $x_0 \wedge b = o$. Since $x_0 \vee b = a$, $x_0 \wedge b = o$ means the same thing as $a = x_0 \otimes b$. According to the induction hypothesis, $d(b) = \Sigma_{0<i<n} d(x_i)$ is equivalent to $b = x_1 \otimes \cdots \otimes x_{n-1}$. Consequently, it follows from 2.7 that if $d(a) = \Sigma_{i<n} d(x_i)$, then $a = x_0 \otimes b = x_0 \otimes x_1 \otimes \cdots \otimes x_{n-1}$, and conversely, if $a = x_0 \otimes x_1 \otimes \cdots \otimes x_{n-1}$, then $a = x_0 \otimes b$ and $b = x_1 \otimes \cdots \otimes x_{n-1}$, so that $d(a) = \Sigma_{i<n} d(x_i)$.

4.2 COROLLARY. If $a = x \otimes \bar{x} = y \otimes \bar{y}$ and $x \vee \bar{y} = y \vee \bar{x} = a$, then $a = y \otimes \bar{x} = x \otimes \bar{y}$.

Proof. By 4.1, $d(a) = d(x \vee \bar{y}) = d(x) + d(\bar{y}) - d(x \wedge \bar{y}) = d(y) + d(\bar{y})$. Thus, $d(x) \geqq d(y)$. Similarly, $d(y) \geqq d(x)$. Hence, $d(x) = d(y)$, and

$d(x \vee \bar{y}) = d(a) = d(y) + d(\bar{y}) = d(x) + d(\bar{y})$. Thus, $a = x \otimes \bar{y}$. By symmetry, $a = y \otimes \bar{x}$.

4.3 LEMMA. If $\{x_0, x_1, \cdots, x_{n-1}\}$ is independent and $0 \leqq y_i \leqq x_i$ for each $i < n$, then $\{y_0, y_1, \cdots, y_{n-1}\}$ is independent.

Proof. $y_i \wedge \bigvee_{j \neq i} y_j \leqq x_i \wedge \bigvee_{j \neq i} x_j = o$.

4.4 LEMMA. If $x \leqq a \leqq x \otimes y$, then $a = x \otimes (y \wedge a)$.

Proof. By modularity, $x \vee (y \wedge a) = (x \vee y) \wedge a = a$. The set $\{x, y \wedge a\}$ is independent by 4.3.

4.5 LEMMA. If $\{x_0, x_1, \cdots, x_{n-1}\}$ is independent in **L**, and if $a \in L$, then

$$\bigvee_{i<n}((a \vee \bar{x}_i) \wedge x_i) = (\bigwedge_{i<n}(a \vee \bar{x}_i)) \wedge (\bigvee_{i<n} x_i),$$

where $\bar{x}_i = \bigvee_{j \neq i} x_j$ for $i < n$.

Proof. If $n = 1$, the result is trivial. Assume that the identity holds for the independent set $\{x_1, \cdots, x_{n-1}\}$ of $n - 1$ elements, and for every $a \in L$. For $1 \leqq i < n$, denote $\tilde{x}_i = x_1 \vee \cdots \vee x_{i-1} \vee x_{i+1} \vee \cdots \vee x_{n-1}$. By the induction hypothesis, $\bigvee_{0<i<n}((a \vee \bar{x}_i) \wedge x_i) = \bigvee_{0<i<n}(((a \vee x_0) \vee \tilde{x}_i) \wedge x_i) = (\bigwedge_{0<i<n}((a \vee x_0) \vee \tilde{x}_i)) \wedge (\bigvee_{0<i<n} x_i) = (\bigwedge_{0<i<n}(a \vee \bar{x}_i)) \wedge \bar{x}_0$. Consequently, two applications of the modular law give

$$\begin{aligned} \bigvee_{i<n}((a \vee \bar{x}_i) \wedge x_i) &= ((a \vee \bar{x}_0) \wedge x_0) \vee \bigvee_{0<i<n}((a \vee \bar{x}_i) \wedge x_i) \\ &= ((a \vee \bar{x}_0) \wedge x_0) \vee ((\bigwedge_{0<i<n}(a \vee \bar{x}_i)) \wedge \bar{x}_0) \\ &= (a \vee \bar{x}_0) \wedge (x_0 \vee ((\bigwedge_{0<i<n}(a \vee \bar{x}_i)) \wedge \bar{x}_0)) \\ &= (a \vee \bar{x}_0) \wedge ((\bigwedge_{0<i<n}(a \vee \bar{x}_i)) \wedge (x_0 \vee \bar{x}_0) \\ &= (\bigwedge_{i<n}(a \vee \bar{x}_i)) \wedge (\bigvee_{i<n} x_i). \end{aligned}$$

This completes the induction.

The preparations for Ore's theorem are completed. The proof of this theorem is somewhat technical. It will be helpful in understanding this proof to think through the entire argument for the case in which **L** is the lattice of all subspaces of a finite dimensional vector space. Of course, the statements of 4.1 through 4.5 should be considered in this light also. In the case of the lattice of subspaces of a vector space, the indecomposable elements are precisely the one-dimensional subspaces, and the concept of independence for a set of such indecomposables coincides with the usual notion of linear independence. Moreover, the expressions of the form $(a \vee \bar{x}_i) \wedge x_i$, which occur in 4.5 and in the proof of Ore's theorem, can be roughly interpreted as the projection of the space a in the space x_i.

4.6 THEOREM. (*Ore*) Let **L** be a finite dimensional lattice. Suppose that $a = x_0 \otimes x_1 \otimes \cdots \otimes x_{m-1} = y_0 \otimes y_1 \otimes \cdots \otimes y_{n-1}$, where the x_i and y_j are indecomposable. Then there is a one-to-one correspondence $i \rightarrow j(i)$

between m and n such that for each $r < m$

$$a = y_{j(0)} \otimes \cdots \otimes y_{j(r)} \otimes x_{r+1} \otimes \cdots \otimes x_{m-1}$$
$$= y_0 \otimes \cdots \otimes y_{j(r)-1} \otimes x_r \otimes y_{j(r)+1} \otimes \cdots \otimes y_{n-1}.$$

In particular $m = n$, and for each r the quotient lattices x_r/o and $y_{j(r)}/o$ are isomorphic.

Proof. It is convenient to introduce the notation

$$\bar{x}_i = \bigvee_{j \neq i} x_j, \qquad \bar{y}_i = \bigvee_{j \neq i} y_i.$$

The case $r = 0$ of the theorem can then be formulated as follows: there exists j such that $a = y_j \otimes \bar{x}_0 = x_0 \otimes \bar{y}_j$. Most of the proof of Ore's theorem is devoted to establishing this fact. We will see that once this result has been proved, the full theorem is easily obtained. The proof is carried out by induction on $d(a)$. The case $d(a) = 0$ is trivial, since $d(a) = 0$ implies $m = 0$. The theorem can therefore be assumed to hold when a is replaced by any $b \in L$ such that $d(b) < d(a)$. The proof is in three steps. The first and most difficult step is to show that if $x_0 \vee \bar{y}_k < a$ for some k, then there exists $j \neq k$ such that

$$a = y_j \otimes \bar{x}_0 = x_0 \otimes \bar{y}_j.$$

The strict inequality $x_0 \vee \bar{y}_k < a$ is what we need to apply the induction hypothesis. Step two of the proof is devoted to the removal of this restriction. The last part of the argument takes us from the case $r = 0$ to the general statement of the theorem.

(i) Assume for the moment that $x_0 \vee \bar{y}_k < a$. Let $z_i = (x_0 \vee \bar{y}_i) \wedge y_i$ for all $i < n$. Define

$$b = z_0 \vee z_1 \cdots \vee z_{n-1} = z_0 \otimes z_1 \otimes \cdots \otimes z_{n-1}.$$

The fact that the join is direct follows from 4.3. By 4.5

$$x_0 \leqq b \leqq a = x_0 \otimes \bar{x}_0.$$

Therefore, by 4.4

$$b = x_0 \otimes (b \wedge \bar{x}_0).$$

We wish to apply the induction hypothesis to b, so that it is necessary to show that $d(b) < d(a)$. By assumption, $x_0 \vee \bar{y}_k < a$. Therefore, $z_k = (x_0 \vee \bar{y}_k) \wedge y_k < y_k$, since otherwise $x_0 \vee \bar{y}_k \geqq y_k \vee \bar{y}_k = a$. By condition (ii) in Definition 3.1 of a dimension function, $z_k < y_k$ implies $d(z_k) < d(y_k)$. Therefore, by 4.1,

$$d(b) = d(z_0) + d(z_1) + \cdots + d(z_{n-1}) < d(y_0) + d(y_1) + \cdots + d(y_{n-1}) = d(a).$$

Consequently, the induction hypothesis is applicable to the element b. By 2.9 and 3.2, every z_i different from o can be written in the form

$$z_i = w_{i0} \otimes \cdots \otimes w_{is(i)},$$

where the w's are indecomposable. Hence,

$$b = w_{00} \otimes \cdots \otimes w_{ji} \otimes \cdots \otimes w_{n-1s} = x_0 \otimes (b \wedge \bar{x}_0).$$

By the induction hypothesis

$$b = w \otimes (b \wedge \bar{x}_0),$$

where

$$w = w_{ji} \leqq z_j \leqq y_j.$$

Consequently,

$$d(b) = d(w) + d(b \wedge \bar{x}_0) = d(w) + d(b) + d(\bar{x}_0) - d(b \vee \bar{x}_0),$$

so that

$$d(b \vee \bar{x}_0) = d(w) + d(\bar{x}_0).$$

Moreover, $a \geqq w \vee \bar{x}_0 = w \vee (b \wedge \bar{x}_0) \vee \bar{x}_0 = b \vee \bar{x}_0 \geqq x_0 \vee \bar{x}_0 = a$. Thus, $a = w \vee \bar{x}_0 = b \vee \bar{x}_0$, and $d(w \vee \bar{x}_0) = d(b \vee \bar{x}_0) = d(w) + d(\bar{x}_0)$. Therefore, by 4.1,

$$a = w \otimes \bar{x}_0.$$

Since $w \leqq y_j \leqq a = w \otimes \bar{x}_0$, Lemma 4.4 yields

$$y_j = w \otimes (y_j \wedge \bar{x}_0).$$

However, by assumption y_j is indecomposable, and since $w > 0$, it follows that

$$w = y_j.$$

Therefore,

$$a = y_j \otimes \bar{x}_0.$$

Moreover, $y_j = w \leqq z_j \leqq y_j$ implies $y_j = z_j = (x_0 \vee \bar{y}_j) \wedge y_j$. Thus

$$a \geqq x_0 \vee \bar{y}_j \geqq y_j \vee \bar{y}_j = a.$$

From the equalities

$$a = x_0 \otimes \bar{x}_0 = y_j \otimes \bar{y}_j = y_j \vee \bar{x}_0 = x_0 \vee \bar{y}_j,$$

it follows by 4.2 that

$$a = x_0 \otimes \bar{y}_j.$$

Finally, note that $j \neq k$, since $x_0 \vee \bar{y}_j = a$ and $x_0 \vee \bar{y}_k < a$. Consequently, the proof of step (i) is complete. Note that the arguments remain valid when x_0 is replaced by any x_i, and also when the roles of the x's and y's are interchanged.

(ii) Next it will be shown that the result obtained in (i) holds without the hypothesis $x_0 \vee \bar{y}_k < a$. That is, there must be some $j < n$ such that $a = y_j \otimes \bar{x}_0 = x_0 \otimes \bar{y}_j$. If $x_0 \vee \bar{y}_k < a$ for some k, then this conclusion follows from (i). If $a = y_k \vee \bar{x}_0 = x_0 \vee \bar{y}_k$ for some k, then $a = y_k \otimes \bar{x}_0 = x_0 \otimes \bar{y}_k$ by 4.2. The only remaining possibility is that $x_0 \vee \bar{y}_k = a$ and $y_k \vee \bar{x}_0 < a$ for all $k < n$. However, it is not hard to see that this case cannot occur. In fact, if $y_k \vee \bar{x}_0 < a$ for all $k < n$, then by applying (i) repeatedly (with x's and y's interchanged), each y_k in the expression $a = y_0 \otimes y_1 \otimes \cdots \otimes y_{n-1}$ can be replaced in its turn by some x_j with $j \neq 0$. The final result is

$$a = x_{j(0)} \otimes x_{j(1)} \otimes \cdots \otimes x_{j(n)} \leqq \bar{x}_0.$$

However, this is a contradiction, since $x_0 \wedge \bar{x}_0 = 0$ and $0 < x_0 \leqq a$.
(iii) The proof of Ore's theorem can now be completed easily. By (ii), there exists $j(0) < n$ such that

$$a = y_{j(0)} \otimes x_1 \otimes \cdots \otimes x_{m-1} = x_0 \otimes \bar{y}_{j(0)}.$$

Applying (ii) again to

$$a = y_{j(0)} \otimes x_1 \otimes \cdots \otimes x_{m-1} = y_0 \otimes y_1 \otimes \cdots \otimes y_{m-1},$$

with 0 replaced by 1, we obtain $j(1) < n$ such that

$$a = y_{j(0)} \otimes y_{j(1)} \otimes x_2 \otimes \cdots \otimes x_{m-1} = x_1 \otimes \bar{y}_{j(1)}.$$

Continuing in this way, we obtain for each $r < m$

$$a = y_{j(0)} \otimes \cdots \otimes y_{j(r)} \otimes x_{r+1} \otimes \cdots \otimes x_{m-1} = x_r \otimes \bar{y}_{j(r)}.$$

Finally,

$$a = y_{j(0)} \otimes y_{j(1)} \otimes \cdots \otimes y_{j(m-1)}.$$

It follows that if $i \neq k$, then $j(i) \neq j(k)$, since otherwise $\{y_{j(0)}, y_{j(1)}, \cdots, y_{j(m-1)}\}$ would not be independent. Hence j maps m one-to-one into n. In particular, $m \leqq n$. By symmetry, $n \leqq m$. Hence, j is one-to-one and onto. Since $a = x_r \otimes \bar{y}_{j(r)} = y_{j(r)} \otimes \bar{y}_{j(r)}$, it follows that x_r/o and $y_{j(r)}/o$ are both transposes of $a/\bar{y}_{j(r)}$. Consequently, these quotient lattices are isomorphic by the transposition principle.

For the application of Ore's theorem, an elementary isomorphism theorem is needed. An algebra **A** is said to have a *one-element subalgebra* if there is a set $\{e\}$ that determines a subalgebra of **A**. In this case, $\{e\}$ is called a one-element subalgebra of **A**. For example, groups, rings, and modules are algebras with one-element subalgebras. All lattices have one-element subalgebras. In fact, every element of the lattice determines a subalgebra of the lattice.

4.7 LEMMA. Let $\mathbf{A} = \langle A; F_\xi \rangle_{\xi < \rho}$ be an algebra of type τ with a one-element subalgebra $\{e\}$. Suppose that $\Delta \in \Theta(\mathbf{A})$. Let $S_\Delta = \{b \in A \mid \langle b, e \rangle \in \Delta\}$. Then $S_\Delta \in \Sigma(\mathbf{A})$. Moreover, if $\Gamma \in \Theta(\mathbf{A})$ is such that $\Gamma \circ \Delta = U_A$ and $\Gamma \cap \Delta = I_A$, then $\mathbf{A}/\Gamma$ is isomorphic to $\mathbf{A} \restriction S_\Delta$.

Proof. The fact that $\{e\}$ is a one-element subalgebra of **A** means that if $\mathbf{e} \in A^{\tau(\xi)}$ is such that $\mathbf{e}(\eta) = e$ for all $\eta < \tau(\xi)$, then $F_\xi(\mathbf{e}) = e$. From this, it follows easily that $S_\Delta \in \Sigma(\mathbf{A})$. Indeed, $\mathbf{a} \in S_\Delta$ implies $\langle \mathbf{a}(\eta), \mathbf{e}(\eta) \rangle \in \Delta$ for all $\eta < \tau(\xi)$. Hence, $\langle F_\xi(\mathbf{a}), e \rangle = \langle F_\xi(\mathbf{a}), F_\xi(\mathbf{e}) \rangle \in \Delta$, since Δ is a congruence relation. Therefore, $F_\xi(\mathbf{a}) \in S_\Delta$. To prove the last statement, observe that if $a \in A$, then $\langle a, e\rangle \in \Gamma \circ \Delta$. Hence, $b \in A$ exists satisfying

$$\langle a, b\rangle \in \Gamma, \qquad \langle b, e\rangle \in \Delta.$$

Moreover, this b is unique, since if also $\langle a, b'\rangle \in \Gamma$ and $\langle b', e\rangle \in \Delta$, then $\langle b, b'\rangle \in \Gamma$ and $\langle b, b'\rangle \in \Delta$. Therefore, $\langle b, b'\rangle \in \Gamma \cap \Delta = I_A$, and $b = b'$. Clearly, $b \in S_\Delta$. Define the mapping φ of A to S_Δ by $\varphi(a) = b$ if $\langle a, b\rangle \in \Gamma$ and $\langle b, e\rangle \in \Delta$. If $\mathbf{a} \in A^{\tau(\xi)}$, then $\varphi \circ \mathbf{a}$ satisfies $\langle \mathbf{a}(\eta), \varphi \circ \mathbf{a}(\eta)\rangle \in \Gamma$, $\langle \varphi \circ \mathbf{a}(\eta), \mathbf{e}(\eta)\rangle \in \Delta$ for all $\eta < \tau(\xi)$. Therefore, $\langle F_\xi(\mathbf{a}), F_\xi(\varphi \circ \mathbf{a})\rangle \in \Gamma$ and $\langle F_\xi(\varphi \circ \mathbf{a}), e\rangle = \langle F_\xi(\varphi \circ \mathbf{a}), F_\xi(\mathbf{e})\rangle \in \Delta$. Thus, $\varphi(F_\xi(\mathbf{a})) = F_\xi(\varphi \circ a)$, so that φ is a homomorphism of $\mathbf{A}$ to $\mathbf{A} \restriction S_\Delta$. If $b \in S_\Delta$, then $\langle b, b\rangle \in \Gamma$, $\langle b, e\rangle \in \Delta$. Hence, $\varphi(b) = b$ for all $b \in S_\Delta$. In particular, since $\mathbf{A}$ is an algebra, it follows that φ is an epimorphism. Finally, $\varphi(a_1) = \varphi(a_2)$ if and only if $\langle a_1, \varphi(a_2)\rangle \in \Gamma$. Since $\langle a_2, \varphi(a_2)\rangle \in \Gamma$, this condition is equivalent to $\langle a_1, a_2\rangle = \Gamma$. That is, $\Gamma_\varphi = \Gamma$. Therefore, $\mathbf{A}/\Gamma$ is isomorphic to $\mathbf{A} \restriction S_\Delta$ by 1.3.10.

Another proof of the last part of this lemma, can be given, using 1.6 to identify $\mathbf{A}$ with $\mathbf{A}/\Gamma \times \mathbf{A}/\Delta$. The subalgebra S_Δ is then identified with the set of all elements of the form $\langle \bar{\Gamma}(a), \bar{\Delta}(e)\rangle \in A/\Gamma \times A/\Delta$, from which it is obvious that $\mathbf{A}/\Gamma = \bar{\Gamma}(A)$ is isomorphic to S_Δ.

4.8 THEOREM. Let $\mathbf{A}$ be an algebra with commuting congruence relations such that $\Theta(\mathbf{A})$ satisfies the ascending and descending chain conditions. Suppose that $\varphi:\mathbf{A} \to \Pi_{i<m}\mathbf{A}_i$ and $\psi:\mathbf{A} \to \Pi_{i<n}\mathbf{B}_i$ are two direct decompositions of $\mathbf{A}$ as direct products of indecomposable algebras. Then $m = n$ and there is a one-to-one mapping j of m onto n such that $\Theta(\mathbf{A}_i)$ is isomorphic to $\Theta(\mathbf{B}_{j(i)})$. Moreover, if $\mathbf{A}$ is an algebra with a one element subalgebra, then $\mathbf{A}_i$ is isomorphic to $\mathbf{B}_{j(i)}$ for each i.

Proof. Let Γ^i be the congruence relation on $\mathbf{A}$ which is the kernel of $\pi_i \circ \varphi$. Similarly, let Δ^i be the kernel of $\pi_i \circ \psi$. Define $\Gamma_i = \bigcap_{j\neq i}\Gamma^j$ and $\Delta_i = \bigcap_{j\neq i}\Delta^j$. By 1.4 and 2.5,

$$U_A = \Gamma_0 \otimes \Gamma_1 \otimes \cdots \otimes \Gamma_{m-1} = \Delta_0 \otimes \Delta_1 \otimes \cdots \otimes \Delta_{n-1}.$$

By 2.4, $\bigvee_{j\neq i}\Gamma_j = \Gamma^i$ and $\bigvee_{j\neq i}\Delta_j = \Delta^i$. Since $\mathbf{A}_i$ is isomorphic to $\mathbf{A}/\Gamma^i$ and $\mathbf{B}_i$ is isomorphic to A/Δ^i, it follows from 2.6 that $\Theta(\mathbf{A}_i)$ and $\Theta(\mathbf{B}_i)$ are respectively isomorphic to Γ_i/I_A and Δ_i/I_A (quotient lattices of $\Theta(\mathbf{A})$). In particular, Γ_i and Δ_i are indecomposable elements of $\Theta(\mathbf{A})$. Thus, it follows from Ore's theorem that $m = n$, and there is a one-to-one mapping j of m onto n such that

$$\Gamma_i \otimes \Gamma^i = \Gamma_i \otimes \Delta^{j(i)},$$

and Γ_i/I_A is isomorphic to $\Delta_{j(i)}/I_A$ for all $i < m$. Consequently, $\Theta(\mathbf{A}_i)$ is isomorphic to $\Theta(\mathbf{B}_{j(i)})$ for all $i < m$. Moreover, if $\mathbf{A}$ has a one-element subalgebra, then it follows from 4.7 that

$$\mathbf{A}_i \cong \mathbf{A}/\Gamma^i = \mathbf{S}_{\Gamma_i} \cong \mathbf{A}/\Delta^{j(i)} \cong \mathbf{B}_{j(i)}.$$

where $\mathbf{S}_{\Gamma_i} = \mathbf{A} \restriction S_{\Gamma_i}$.

Problems

1. For each i in an infinite index set I, let $\mathbf{A}_i$ be a copy of the additive group of integers $\langle A_i; +\rangle$. Let $\Sigma_{i\in I}A_i = \{x \in \mathbf{X}_{i\in I}A_i \mid x(i) = 0$ for all but finitely many $i\}$, that is $\Sigma_{i\in I}A_i$ is the direct sum of the set $\{\mathbf{A}_i \mid i \in I\}$ of Abelian groups. Clearly, $\Sigma_{i\in I}A_i$ determines a subalgebra of $\Pi_{i\in I}\mathbf{A}_i$. Let F denote the addition operation of $\Pi_{i\in I}\mathbf{A}_i$ restricted to $\Sigma_{i\in I}A_i$. Let z and w be the elements of $\mathbf{X}_{i\in I}A_i$ defined by $z(i) = 1$, $w(i) = 2$ for all $i \in I$. Let $B = \Sigma_{i\in I}A_i \cup \{z, w\}$ and $\mathbf{B} = \langle B; F\rangle$. Let $\Gamma_i = \{\langle x, y\rangle \in B^2 \mid x(i) = y(i)\}$. Prove that $\Gamma_i \in \Theta(\mathbf{B})$, and $\mathbf{B}/\Gamma_i$ is isomorphic to $\mathbf{A}_i$. Show that $\{\Gamma_i \mid i \in I\}$ determines an interdirect decomposition $\langle \varphi; \Pi_{i\in I}\mathbf{B}/\Gamma_i\rangle$ (with $\pi_i \circ \varphi = \bar{\Gamma}_i$) for which $\varphi(B)$ is not a subalgebra of $\Pi_{i\in I}\mathbf{B}/\Gamma_i$ and $\mathbf{B}$ is not isomorphic to $(\Pi_{i\in I}\mathbf{B}/\Gamma_i) \restriction \varphi(B)$.

2. Let $\mathbf{A}$ be the additive group of rational numbers. Prove that the lattice $\Theta(\mathbf{A})$ satisfies the restricted chain condition, but that it does not satisfy either the descending chain condition or the ascending chain condition.

3. Let $\mathbf{L}$ be a lattice with a zero element o. A *rank function* r on $\mathbf{L}$ is a mapping of L into the integers such that

$$\begin{aligned}&r(a \wedge b) + r(a \vee b) = r(a) + r(b),\\ &a \leqq b \text{ implies } r(a) \leqq r(b), \text{ and}\\ &r(a) = 0 \text{ if and only if } a = o.\end{aligned}$$

(a) Prove that if $\mathbf{L}$ is relatively complemented, then every rank function is a dimension function.

(b) Show that if $\mathbf{L}$ is modular and there is a rank function defined on $\mathbf{L}$, then $\mathbf{L}$ satisfies the restricted chain condition.

(c) Let $\mathbf{A}$ be a subgroup of a finite, dimensional, rational vector space. For $B \in \Sigma(\mathbf{A})$, let $r(B)$ be the cardinal number of a maximal linearly independent subset of B. Prove that r is a rank function on $\Sigma(\mathbf{A})$. Use this fact and the isomorphism between $\Theta(\mathbf{A})$ and $\Sigma(\mathbf{A})$ noted in Example 2.4.4 to conclude that $\mathbf{A}$ can be written as a direct product of indecomposable groups.

4. Let V be the four-dimensional, rational vector space with the basis $\{w, x, y, z\}$. Let $\mathbf{A}$ be the subgroup of $\mathbf{V}$ determined by the set

$$\begin{aligned}A = \{&(a/5^n)w + (b/5^n)x + (c/7^n)y + (d/11^n)z + (e/3)(x + y)\\ &+ (f/2)(x + z) \mid a, b, c, d, e, f \text{ integers}, n < \omega\}.\end{aligned}$$

Let $\mathbf{B}$, $\mathbf{C}$, $\mathbf{D}$, and $\mathbf{E}$ be the subgroups of $\mathbf{A}$ determined by the respective sets

$$\begin{aligned}B &= \{(a/5^n)w \mid a \text{ an integer}, n < \omega\}.\\ C &= \{(a/5^n)x + (b/7^n)y + (c/11^n)z + (d/3)(x + y)\\ &\qquad + (e/2)(x + z) \mid a, b, c, d, e \text{ integers}, n < \omega\}.\end{aligned}$$

$D = \{(a/5^n)(3w - x) + (b/7^n)y + (c/3)(3w - x - y) \mid a, b, c$ integers, $n < \omega\}$.

$E = \{(a/5^n)(2w - x) + (b/11^n)z + (c/2)(2w - x - z) \mid a, b, c$ integers, $n < \omega\}$.

(a) Prove that $B \cap C = D \cap E = \{0\}$, and (in $\Sigma(\mathbf{A})$) $B \vee C = D \vee E = A$. Deduce that **A** is isomorphic to the direct product of **B** and **C** and also to the direct product of **D** and **E**. (*Hint:* show that the mapping of $B \times C$ to A given by $\langle b, c\rangle \to b + c$ is an isomorphism of $\mathbf{B} \times \mathbf{C}$ to **A**.)

(b) Show that neither **B** nor **C** is isomorphic to either of **D** or **E**. (*Hint:* consider the ranks of these groups—see Problem **3(c)**.)

(c) Prove that **B**, **D**, and **E** are indecomposable. (*Hint:* suppose that $\{\Gamma, \Delta\} \subseteq \Theta(\mathbf{D})$ determines a proper direct decomposition of **D**. Define $D_\Gamma = \{u \in D \mid \langle u, 0\rangle \in \Gamma\}$, $D_\Delta = \{u \in D \mid \langle u, 0\rangle \in \Delta\}$. It is easily seen that D_Γ and D_Δ determine subgroups of **D**. Note also that $D_\Gamma \cap D_\Delta = \{0\}$. Consequently, if $u \in D_\Gamma$, then no nonzero rational multiple of u lies in D_Δ and vice versa. Moreover, for each $u \in D$, there exists a unique $v \in D_\Delta$ such that $u - v \in D_\Gamma$ and a unique $w \in D_\Gamma$ such that $u - w \in D_\Delta$. Using these facts with u successively taken to be $y, 7^{-1}y, 7^{-2}y, \cdots$, we see that there exists $v_n \in D_\Delta$ such that $7^{-n}y - v_n \in D_\Gamma$ for all n. From uniqueness $v_n = 7^{-n}v_0$. From the fact that $7^{-n}v_0 \in D_\Delta \subseteq D$ for all n, it follows that either v_0 is 0 (so that $7^{-n}y \in D_\Gamma$), or else v_0 is a nonzero rational multiple of y. In this last case $v_0 = y$, since otherwise D_Γ and D_Δ would both contain nonzero rational multiples of y. It can be supposed therefore that $7^{-n}y \in D_\Gamma$ for all n. Then it follows easily that D_Γ contains only integral multiples of $7^{-n}y$, $n < \omega$, because otherwise D_Γ would contain a multiple of every element of D, and hence $D_\Delta = \{0\}$. Similarly, D_Δ consists of the integral multiples $5^{-n}(3w - x)$, $n < \omega$. This yields the final contradiction, since it will be impossible to find $v \in D_\Gamma$ such that $(1/3)(3w - x - y) - v \in D_\Delta$.)

From the results of **(a)**, **(b)**, and **(c)** it follows that **A** has two essentially different decompositions as a direct product of indecomposables. Thus, there is no chance of proving a version of Ore's theorem that does not have a chain condition in the hypothesis.

5. Let **A** be a partial algebra, and $\{\Gamma_i \mid i \in I\} \subseteq \Theta(\mathbf{A})$. Suppose that $\{\Gamma_i \mid i \in I\}$ satisfies the condition of Proposition 1.5 (so that this set determines an interdirect decomposition of **A**). Use the result of Proposition 1.5 to prove that

$$(\textstyle\bigcap_{i\in I-F}\Gamma_i) \circ (\bigcap_{i\in I-F}\Gamma_i) = \bigcap_{i\in I-(F\cup G)}\Gamma_i$$

for any two finite sets F and G in I.

This result generally fails in the case that F and G are infinite (see Problem **11** below).

6. **(a)** Let Γ and Δ be equivalence relations on a set A. Prove that

$$\Gamma \cup (\Gamma \circ \Delta) \cup (\Gamma \circ \Delta \circ \Gamma) \cup (\Gamma \circ \Delta \circ \Gamma \circ \Delta) \cup \cdots$$

is an equivalence relation on A which is the least upper bound of Γ and Δ in $E(A)$.

(b) Let $\mathbf{A}$ be a finitary algebra. Prove that $\Theta(\mathbf{A})$ is a sublattice of $E(A)$. (*Hint:* this will follow easily from the result that if Γ and Δ are congruence relations, then so is $\Gamma \cup (\Gamma \circ \Delta) \cup (\Gamma \circ \Delta \circ \Gamma) \cup (\Gamma \circ \Delta \circ \Gamma \circ \Delta) \cup \cdots$.)

(c) Use the result of Problem **7(b)** in Chapter 2 to show that there exists a finitary partial algebra $\mathbf{A}$ for which $\Theta(\mathbf{A})$ is not a sublattice of $E(A)$.

(d) Let $\mathbf{A} = \langle \omega + 1; F \rangle$, where F is the ω-ary operation defined by $F(\mathbf{x}) = \sup \{\mathbf{x}(i) \mid i < \omega\}$. Show that

$$\Gamma = \{\langle 2i, 2i+1 \rangle \mid i < \omega\} \cup \{\langle 2i+1, 2i \rangle \mid i < \omega\} \cup I_{\omega+1}$$

and

$$\Delta = \{\langle 2i-1, 2i \rangle \mid i < \omega\} \cup \{\langle 2i, 2i-1 \rangle \mid i < \omega\} \cup I_{\omega+1}$$

are congruence relations on $\mathbf{A}$, but that the join of Γ and Δ in $E(A)$ is not a congruence relation on $\mathbf{A}$.

(e) Let $\mathbf{A}$ be a finitary partial algebra. Suppose that $\mathfrak{D} \subseteq \Theta(\mathbf{A})$, where $\mathfrak{D} \neq \varnothing$. Prove that the least upper bound of $\mathfrak{D}$ in $\Theta(\mathbf{A})$ is $\bigcup\{\Delta_0 \vee \Delta_1 \vee \cdots \vee \Delta_{n-1} \mid \Delta_i \in \mathfrak{D}, n = 1, 2, \cdots\}$ (where $\Delta_0 \vee \Delta_1 \vee \cdots \vee \Delta_{n-1}$ denotes the join in $\Theta(\mathbf{A})$).

(f) Use the result of **(b)** and **(e)** to prove that if $\mathbf{A}$ is a finitary algebra, and $\mathfrak{D}$ is a nonempty subset of $\Theta(\mathbf{A})$, then the least upper bound of $\mathfrak{D}$ in $\Theta(\mathbf{A})$ is

$$\bigcup\{\Delta_0 \circ \Delta_1 \circ \cdots \circ \Delta_{n-1} \mid \Delta_i \in \mathfrak{D}, n = 1, 2, \cdots\}.$$

(g) With the same hypotheses as in **(f)**, show that if $\Gamma \in \Theta(\mathbf{A})$ and if Γ commutes with all $\Delta \in \mathfrak{D} \subseteq \Theta(\mathbf{A})$, then Γ commutes with the least upper bound of $\mathfrak{D}$ in $\Theta(\mathbf{A})$.

7. Let $\mathbf{L}$ be a lattice. Let $\Gamma \in \Theta(\mathbf{L})$, $\mathfrak{D} \subseteq \Theta(\mathbf{L})$, $\mathfrak{D} \neq \varnothing$.

(a) Prove that if $a \leqq x \leqq b$ and $a \leqq y \leqq b$, and if $\langle a, b \rangle \in \Gamma$, then $\langle x, y \rangle \in \Gamma$.

(b) Prove that $\langle x, y \rangle \in \Gamma$ if and only if $\langle x \wedge y, x \vee y \rangle \in \Gamma$.

(c) Let $a \leqq b$. Prove that $\langle a, b \rangle \in \bigvee\mathfrak{D}$ if and only if there exist $\Delta_0, \Delta_1, \cdots, \Delta_{n-1}$ in $\mathfrak{D}$, and $y_0, y_1, \cdots, y_{n-1}, y_n$ in L such that $a = y_0 \leqq y_1 \leqq \cdots \leqq y_{n-1} \leqq y_n = b$, and $\langle y_i, y_{i+1} \rangle \in \Delta_i$ for $i < n$. (*Hint:* by **6(f)**, $\Delta_0, \Delta_1, \cdots, \Delta_{n-1}$ exist in $\mathfrak{D}$ such that $\langle a, b \rangle \in \Delta_0 \circ \Delta_1 \circ \cdots \circ \Delta_{n-1}$. Consequently, $x_0 = a, x_1, \cdots, x_{n-1}, x_n = b$ exist in L such that $\langle x_i, x_{i+1} \rangle \in \Delta_i$. Let $y_i = (x_0 \vee x_1 \vee \cdots \vee x_i) \wedge b$.)

(d) Prove that $\Gamma \cap \bigvee\mathfrak{D} \subseteq \bigvee\{\Gamma \cap \Delta \mid \Delta \in \mathfrak{D}\}$. (*Hint:* let $\langle x, y \rangle \in \Gamma \cap \bigvee\mathfrak{D}$. Use **(a)**, **(b)**, and **(c)** with $a = x \wedge y$, $b = z \vee y$ to show that $\langle x, y \rangle \in \bigvee\{\Gamma \cap \Delta \mid \Delta \in \mathfrak{D}\}$.) Since it is obvious that $\Gamma \cap \bigvee\mathfrak{D} \supseteq \bigvee\{\Gamma \cap \Delta \mid \Delta \in \mathfrak{D}\}$, this result proves that the lattice $\Theta(\mathbf{L})$ satisfies the infinite distributive law

$$\Gamma \cap \bigvee\mathfrak{D} = \bigvee\{\Gamma \cap \Delta \mid \Delta \in \mathfrak{D}\}.$$

In particular, if $\mathfrak{D} = \{\Delta, E\}$, then

$$\Gamma \cap (\Delta \vee E) = (\Gamma \cap \Delta) \vee (\Gamma \cap E).$$

From this distributive law, the dual follows: $(\Gamma \vee \Delta) \cap (\Gamma \vee E) = ((\Gamma \vee \Delta) \cap \Gamma) \vee ((\Gamma \vee \Delta) \cap E) = \Gamma \vee ((\Gamma \vee \Delta) \cap E) = \Gamma \vee ((\Gamma \cap E) \vee (\Delta \cap E)) = (\Gamma \vee (\Gamma \cap E)) \vee (\Delta \cap E) = \Gamma \vee (\Delta \cap E)$. Thus, for any lattice **L**, $\Theta(\mathbf{L})$ is a distributive lattice. The validity of the infinite distributive law $\Gamma \cap \bigvee\mathfrak{D} = \bigvee\{\Gamma \cap \Delta \mid \Delta \in \mathfrak{D}\}$ is of some interest. However, it is not hard to see that every compactly generated distributive lattice satisfies the infinite distributive law

$$x \wedge \bigvee_{i\in I} y_i = \bigvee_{i\in I}(x \wedge y_i).$$

8. **(a)** Let $\mathbf{A} = \langle A; F_\xi\rangle_{\xi<\rho}$ be a partial algebra. Any partial algebra of the form $\langle A; F_\xi\rangle_{\xi\in J}$, where $J \subseteq \rho$ is called a *reduct* of **A**. Prove that if **A** is a finitary algebra, and $\mathbf{A}'$ is a reduct of **A**, then $\Theta(\mathbf{A})$ is a sublattice of $\Theta(\mathbf{A}')$. (See Problem **9**, in Chapter 1.)

(b) Prove that every lattice-ordered group (see Problem **14** in Chapter 2) **A** has commuting congruence relations, and $\Theta(\mathbf{A})$ is distributive.

9. Let $\mathbf{A}_0 = \langle A_0; +, F\rangle$, $\mathbf{A}_1 = \langle A_1; +, G\rangle$ be abstract algebras of type $\{\langle 0, 2\rangle, \langle 1, 1\rangle\}$ defined as follows: $A_0 = A_1 = \{0, 1\}$; $0 + 0 = 1 + 1 = 0$, $0 + 1 = 1 + 0 = 1$ (thus, $\langle A_0; +\rangle$ and $\langle A_1; +\rangle$ are cyclic groups of order two); $F(0) = 1$, $F(1) = 0$; $G(0) = 0$, $G(1) = 1$. Define **A** to be the direct product of $\{\mathbf{A}_0, \mathbf{A}_1\}$.

(a) Prove that $\mathbf{A}_0$ and $\mathbf{A}_1$ are indecomposable, and that $\mathbf{A}_0$ is not isomorphic to $\mathbf{A}_1$.

(b) Show that **A** has commuting congruence relations, and $\Theta(\mathbf{A})$ satisfies the ascending and descending chain conditions.

(c) Prove that **A** is isomorphic to $(\mathbf{A}_0)^2$, that is, to the direct product of $\mathbf{A}_0$ with itself.

This example shows that the last statement of Theorem 4.8 may not be true in the absence of the assumption that **A** has a one element subalgebra.

10. Let **A** be an algebra such that $\Theta(\mathbf{A})$ is distributive. Let $\langle\varphi; \Pi_{i\in I}\mathbf{A}/\Gamma_i\rangle$ and $\langle\psi; \Pi_{j\in J}\mathbf{A}/\Delta_j\rangle$ be direct product decompositions of **A** such that $\pi_i \circ \varphi = \bar{\Gamma}_i$ and $\pi_j \circ \psi = \bar{\Delta}_j$. Define $E_{ij} = \Gamma_i \vee \Delta_j$. Let $\Gamma_i' = \bigcap_{k\neq i}\Gamma_k$ and $\Delta_j' = \bigcap_{l\neq j}\Delta_l$.

(a) Prove: $\Gamma_i \vee \Gamma_i' = \Delta_j \vee \Delta_j' = U_A$ and $\Gamma_i \cap \Gamma_i' = \Delta_j \cap \Delta_j' = I_A$ for all $i \in I$ and $j \in J$.

(b) Show that $\Gamma_i' \cap E_{ij} = \Gamma_i' \cap \Delta_j$ for all $i \in I$ and $j \in J$.

(c) Deduce from **(b)** that $\Gamma_i' \cap \bigcap_{j\in J}E_{ij} = I_A$.

(d) Use the result of **(c)** to prove that $\bigcap_{j\in J}E_{ij} = \Gamma_i$. (*Hint:* apply the distributive law to $\bigcap_{j\in J}E_{ij} = (\bigcap_{j\in I}E_{ij}) \cap (\Gamma_i' \vee \Gamma_i)$. Note that $\bigcap_{j\in I}E_{ij} \supseteq \Gamma_i$ by the definition of E_{ij}.)

(e) Prove that there is a monomorphism

$$\chi_i : \mathbf{A}/\Gamma^i \to \Pi_{j\in J}\mathbf{A}/E_{ij}$$

such that the diagram

$$\begin{array}{ccc} \mathbf{A} & \xrightarrow{\quad \bar{\Gamma}_i \quad} & \mathbf{A}/\Gamma_i \\ & {\scriptstyle\vartheta_i}\searrow \quad \swarrow{\scriptstyle\chi_i} & \\ & \Pi_{j\in J}\mathbf{A}/E_{ij} & \end{array}$$

is consistent, where ϑ_i is the unique homomorphism of $\mathbf{A}$ to $\Pi_{j\in J}\mathbf{A}/E_{ij}$ satisfying $\pi_{ij} \circ \vartheta_i = \bar{E}_{ij}$ (see 1.5.6).

(f) Prove that χ_i is an isomorphism. (*Hint:* it is sufficient to show that ϑ_i is an epimorphism. If $x \in \mathbf{X}_{j\in J}A/E_{ij}$, it is possible to choose elements $x_j \in A$ such that $\pi_{ij}(x) = \bar{E}_{ij}(x_j)$. Then since $\langle\psi; \Pi_{j\in J}\mathbf{A}/\Delta_j\rangle$ is a direct product decomposition, there exists $y \in A$ satisfying $\bar{\Delta}_j(y) = \bar{\Delta}_j(x_j)$ for all $j \in J$. Show that $\vartheta_i(y) = x$.)

(g) Prove that $\mathbf{A} \cong \Pi_{\langle i,j\rangle\in I\times J}\mathbf{A}/E_{ij}$, $\mathbf{A}/\Gamma_i \cong \Pi_{j\in J}\mathbf{A}/E_{ij}$, and $\mathbf{A}/\Delta_j \cong \Pi_{i\in I}\mathbf{A}/E_{ij}$. Use this result to show that if $\mathbf{A}$ is decomposed as a direct product of indecomposable algebras in two different ways

$$\langle\varphi; \Pi_{i\in I}\mathbf{A}_i\rangle, \qquad \langle\psi; \Sigma_{j\in J}\mathbf{B}_j\rangle$$

(with all $\mathbf{A}_i$ and $\mathbf{B}_j$ indecomposable algebras, each one containing at least two elements), then there is a one-to-one correspondence λ between I and J, and corresponding isomorphisms $\sigma_i : \mathbf{A}_i \to \mathbf{B}_{\lambda(i)}$ such that $\sigma_i \circ \pi_i \circ \varphi = \pi_{\lambda(i)} \circ \psi$ for all i.

11. Let $\mathbf{A}$ be an algebra. A congruence relation $\Gamma \in \Theta(\mathbf{A})$ is called a *decomposition congruence* if there is a congruence relation Γ' in $\Theta(\mathbf{A})$ such that $\{\Gamma, \Gamma'\}$ determines a decomposition of $\mathbf{A}$ as a direct product of $\{\mathbf{A}/\Gamma, \mathbf{A}/\Gamma'\}$. Thus, by the Propositions 1.4 and 1.5, Γ is a decomposition congruence if and only if there is a congruence Γ' such that $\Gamma \cap \Gamma' = I_A$, $\Gamma \circ \Gamma' = U_A$.

(a) Let $\mathbf{A}$ be an algebra such that $\Theta(\mathbf{A})$ is distributive. Prove that a congruence Γ on $\mathbf{A}$ is a decomposition congruence if and only if Γ has a complement in the lattice $\Theta(\mathbf{A})$ (that is, there exists $\Gamma' \in \Theta(\mathbf{A})$ such that $\Gamma \cap \Gamma' = I_A$ and $\Gamma \vee \Gamma' = U_A$) and $\Gamma \circ \Delta = \Delta \circ \Gamma$ for all $\Delta \in \Theta(\mathbf{A})$. (*Hint:* suppose that $\Gamma \cap \Gamma' = I_A$ and $\Gamma \circ \Gamma' = U_A$. Let $\Delta \in \Theta(\mathbf{A})$ be arbitrary. Suppose that $\langle x, y\rangle \in \Gamma \vee \Delta$. Choose $z \in A$ so that $\langle x, z\rangle \in \Gamma$, $\langle z, y\rangle \in \Gamma'$. Conclude that $\langle z, y\rangle \in \Gamma' \cap (\Gamma \vee \Delta) = \Gamma' \cap \Delta \subseteq \Delta$.)

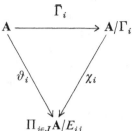

(b) Let $\mathbf{A}$ be an algebra such that $\Theta(\mathbf{A})$ is distributive. Prove that the set $\Theta_0(\mathbf{A})$ of all decomposition congruences determines a sublattice of $\Theta(\mathbf{A})$ which is a Boolean algebra. Show that $\mathbf{A}$ satisfies the restricted chain condition if and only if $\Theta_0(\mathbf{A})$ is a finite Boolean algebra. (*Hint:* use Problem **9** for the first statement. To prove the second statement, it is sufficient to show that in any infinite Boolean algebra there is an infinite sequence $x_0 < x_1 < x_2 < \cdots$.)

12. Let $\mathbf{A}$ and $\{\mathbf{A}_i \mid i \in I\}$ be relational systems of type τ. A *decomposition of* $\mathbf{A}$ *as a direct sum* of $\{\mathbf{A}_i \mid i \in I\}$ is a monomorphism φ of $\mathbf{A}$ to $\Pi_{i\in I}\mathbf{A}_i$ such that

(i) $\langle \varphi; \Pi_{i\in I}\mathbf{A}_i \rangle$ is an interdirect product decomposition of $\mathbf{A}$, and
(ii) if x and y are in A, then $\{i \in I \mid (\pi_i \circ \varphi)(x) \neq (\pi_i \circ \varphi)(y)\}$ is finite.

Note that direct sum decompositions are the minimal interdirect product decompositions. Thus, for finite index sets they coincide with direct product decompositions. In the case of modules, this definition of direct sum decomposition agrees with the usual one. It is not always possible to define the direct sum of an infinite set of algebras. For example, in the class of rings with an identity, no infinite set of nonzero algebras has a direct sum, since the identity and the zero must be different elements at each component.

Let $\{\Gamma_i \mid i \in I\}$ be a set of congruence relations on $\mathbf{A}$ that determines an interdirect decomposition $\langle \varphi; \Pi_{i\in I}\mathbf{A}/\Gamma_i \rangle$ of $\mathbf{A}$. Suppose that $\Gamma_i \neq U_A$ for all $i \in I$. Prove that the following conditions are equivalent:

(1) $\langle \varphi; \Pi_{i\in I}\mathbf{A}/\Gamma_i \rangle$ is a direct sum decomposition:
(2) $U_A = \bigcup\{\bigcap_{i\in I-F}\Gamma_i \mid F \subseteq I,\ F \text{ finite}\}$
(3) The mapping $S \to \Gamma'_S = \bigcap_{i\in I-S}\Gamma_i$ is a complete lattice monomorphism of $\mathbf{P}(I)$ to $\Theta(\mathbf{A})$. That is, if $\{S_k \mid k \in K\} \subseteq \mathbf{P}(I)$ and $S = \bigcap_{k\in K}S_k$, $T = \bigcup_{k\in K}S_k$, then Γ'_S is the greatest lower bound of $\{\Gamma'_{S_k} \mid k \in K\}$ in $\Theta(\mathbf{A})$ and Γ'_T is the least upper bound of $\{\Gamma'_{S_k} \mid k \in K\}$ in $\Theta(\mathbf{A})$. Show that if these equivalent conditions are satisfied, then Γ'_S is a decomposition congruence (see Problem **11**) for all $S \subseteq I$. (*Hint:* note that in any case Γ'_S is the greatest lower bound and Γ'_T is an upper bound of $\{\Gamma'_{S_k} \mid k \in K\}$. Also, by the result of Problem **5**, Γ'_T is the least upper bound of $\{\Gamma'_{S_k} \mid k \in K\}$ when K and all of the sets S_k, $k \in K$, are finite. If (2) holds, it then follows easily—for example, using Problem **4(d)**, Chapter 2 and Problem **6** above—that $S \to \Gamma'_S$ is a complete lattice homomorphism. The converse and the equivalence of (1) and (2) are easily proved.)

The following two problems provide some lattice theoretical results which are useful for the study of direct sum decompositions.

13. A lattice $\mathbf{L}$ is said to be *complemented* if $\mathbf{L}$ has a least element o and a greatest element u, and for each $x \in L$ there exists $x' \in L$ such that $x \wedge x' = o$ and $x \vee x' = u$. The element x' is naturally called the complement of x in L.

(a) Let $\mathbf{L}$ be a complemented modular lattice. Let $a \leqq x \leqq b$ in $\mathbf{L}$. Prove that if x' is a complement of x in $\mathbf{L}$, then $y = a \vee (x' \wedge b)$ is a relative complement of x in b/a. That is, $x \wedge y = a$ and $x \vee y = b$. Hence, every complemented modular lattice is relatively complemented.

(b) Let **L** be a compactly generated, complemented, modular lattice. Prove that every nonzero element of L contains an atom p, i.e., $p/o = \{o, p\}$. (*Hint:* let $o < a$. By Problem **4(f)** in Chapter 2, there exists c and d in a/o such that $c < d$ and $\{x \in L \mid c < x < d\}$ is empty. Let e be a relative complement of c in a/o. Define $p = e \wedge d$. Show that p/o and d/c are transposes of each other, so that p is an atom contained in a.) Prove that every element of L other than u is contained in a dual atom.

(c) Prove that every element of a compactly generated, complemented, modular lattice is a join of atoms and a meet of dual atoms.

(d) Call a set X of atoms independent if $x \wedge \bigvee(X - \{x\}) = o$ for every $x \in X$. Show that the independent sets of atoms satisfy the conditions of Zorn's lemma, so that there exists a maximal independent set of atoms in L. (*Hint:* note that if x is an atom and $x \wedge \bigvee(X - \{x\}) \neq o$, then $x \leqq \bigvee(X - \{x\})$. Moreover, since **L** is compactly generated, the atoms must be compact.)

(e) Prove that if X is a maximal independent set of atoms in L, then $\bigvee X = u$.

14. Let **L** be a complete, upper continuous, modular lattice. (See Problem **4(g)** in Chapter 2 for the definition of upper continuity. Note that any compactly generated lattice is upper continuous.) Suppose that L contains a set $\{x_i \mid i \in I\}$ of atoms such that

(i) $\bigvee_{i \in I} x_i = u$
(ii) $x_i \wedge \bigvee_{j \neq i} x_j = o$ for all $i \in I$.

For $S \subseteq I$, define

$$x'(S) = \bigvee_{i \in S} x_i$$
$$x_i' = x(I - \{i\}).$$

Note that by these definitions and the hypotheses (i) and (ii):

$$x'(I) = u, \qquad \bigvee_{k \in K} x'(S_k) = x'(\textstyle\bigcup_{k \in K} S_k),$$
$$x_i \wedge x_i' = o, \qquad x_i \vee x_i' = u, \qquad x_i \neq o, \qquad x_i' \neq u.$$

(a) Show that $i \in S$ implies $x_i' \vee x'(S) = u$ and $i \notin S$ implies $x_i' \geqq x'(S)$. Thus, $S \neq T$ in $\mathbf{P}(I)$ implies $x'(S) \neq x'(T)$.

(b) Let $a \in L$. Use the upper continuity of **L** to prove that there is a set $T \subseteq I$ that is maximal with the property

$$a \wedge x'(T) = o.$$

Prove that $a \vee x'(T) = u$. (*Hint:* suppose that $i \in I - T$. If $(a \vee x'(T)) \wedge x_i = o$, then $a \wedge x'(T \cup \{i\}) = a \wedge (a \vee x'(T)) \wedge (x_i \vee x'(T)) = a \wedge (((a \vee x(T)) \wedge x_i) \vee x'(T)) = a \wedge x'(T) = o$, which contradicts the maximality of T. Thus, since x_i is an atom, $a \vee x'(T) \geqq x_i$.)

(c) Apply **(b)** with $a = x'(S)$ to prove that $x'(I - S)$ is a complement of $x'(S)$ in L. Hence, the mapping $S \to x'(S)$ is a complete lattice monomorphism of $\mathbf{P}(I)$ to $\mathbf{L}$.

It is an immediate consequence of Problems **13** and **14** that a compactly generated, modular lattice $\mathbf{L}$ contains an independent set $\{x_i \mid i \in I\}$ of atoms such that $\bigvee_{i \in I} x_i = u$ if and only if $\mathbf{L}$ is complemented. Combining this remark with the result of Problem **12**, it is easy to obtain the following characterization of algebras which are direct sums of simple algebras.

15. (a) Let $\mathbf{A}$ be an abstract algebra with commuting congruence relations. Show that $\mathbf{A}$ admits a decomposition as a direct sum of simple algebras if and only if $\Theta(\mathbf{A})$ is complemented. Moreover, if $\mathbf{A}$ has a one-element subalgebra, then every epimorphic image of $\mathbf{A}$ admits a decomposition into a direct sum of simple algebras that are isomorphic to algebras occurring in a factorization of $\mathbf{A}$ as a direct sum of simple algebras. (It can be shown that in this case, direct sum decompositions are unique up to isomorphism.)

(b) A module $\mathbf{A}$ over a ring $\mathbf{R}$ is called *semisimple* if it is a direct sum of simple modules. (Note that this usage of the term semisimple differs from the definition given in Problem **11**, Chapter 2. The term simple has the same meaning as in that problem however.) Use the result of part **(a)** to prove that a module $\mathbf{A}$ is semisimple if and only if every submodule of $\mathbf{A}$ is a direct summand of $\mathbf{A}$.

16. Let $\mathbf{A}$ be an algebra, and let $\Theta_0(\mathbf{A})$ be the set of decomposition congruences of $\mathbf{A}$ (see Problem **11**).

(a) Let $\Delta \in \Theta_0(\mathbf{A})$. Prove that $\mathbf{A}/\Delta$ is directly indecomposable if and only if Δ is a dual atom of $\Theta_0(\mathbf{A})$, that is, $\{\Gamma \in \Theta_0(\mathbf{A}) \mid \Delta \subset \Gamma \subset U_A\} = \varnothing$.

Assume now that if Γ and Δ are in $\Theta_0(\mathbf{A})$, then $\Gamma \circ \Delta = \Delta \circ \Gamma$, and that $\Theta_0(\mathbf{A})$ is a closed sublattice of $\Theta(\mathbf{A})$, that is, $X \subseteq \Theta_0(\mathbf{A})$ implies that the least upper bound and greatest lower bound of X in $\Theta(\mathbf{A})$ belong to $\Theta_0(\mathbf{A})$. Prove that the following results.

(b) $\Theta_0(\mathbf{A})$ is a complemented, modular lattice containing I_A and U_A.

(c) $\Theta_0(\mathbf{A})$ is compactly generated. (See Problem **4(e)**, Chapter 2). Use these facts, together with the result of Problems **12** and **13**, to show that $\mathbf{A}$ has a decomposition as a direct sum of directly indecomposable algebras.

17. Let $\mathbf{A}$ be a finitary algebra such that $\Theta(\mathbf{A})$ is distributive. Let $\Theta_0(\mathbf{A})$ be the sublattice of $\Theta(\mathbf{A})$ consisting of the decomposition congruences of $\mathbf{A}$ (see Problem **11**). Prove that the following conditions are equivalent:

(i) $\mathbf{A}$ admits a decomposition as a direct sum of indecomposable algebras.

(ii) The join (in $\Theta(\mathbf{A})$) of the set of all elements which are atoms of $\Theta_0(\mathbf{A})$ is U_A.

(iii) $\Theta_0(\mathbf{A})$ is a closed sublattice of $\Theta(\mathbf{A})$, that is, the join and meet in $\Theta(\mathbf{A})$ of subsets of $\Theta_0(\mathbf{A})$ belong to $\Theta_0(\mathbf{A})$. (*Hints:* to prove that (i) implies (ii), use Problems **11**, **12**, and **16**, noting that $\Theta_0(\mathbf{A})$ is a Boolean algebra containing I_A and U_A, and that the complement of a dual atom is an atom. The fact that (iii) implies (i) follows from Problems **11** and **16**. To prove that (ii) implies (iii), first note that it is sufficient to show that $\Theta_0(\mathbf{A})$ is closed under joins in $\Theta(\mathbf{A})$. This follows from the fact that $\Theta_0(\mathbf{A})$ is a Boolean algebra that contains the zero and unit of the upper continuous distributive lattice $\Theta(\mathbf{A})$ by proving that if $\mathfrak{D} = \{\Delta_i \mid i \in I\} \subseteq \Theta_0(\mathbf{A})$, then $(\bigvee_{i\in I}\Delta_i')'$ is the greatest lower bound of $\mathfrak{D}$ in $\Theta(\mathbf{A})$. By Problem **6(g)**, the join in $\Theta(\mathbf{A})$ of any subset of $\Theta_0(\mathbf{A})$ commutes with every congruence. Finally, because of the upper continuity, $(\bigvee_{i\in I}\Delta_i) \wedge (\bigwedge_{i\in I}\Delta_i') = I_A$, and it is easy to show that $(\bigvee_{i\in I}\Delta_i) \vee (\bigwedge_{i\in I}\Delta_i')$ contains every atom of $\Theta_0(\mathbf{A})$. The conclusion (iii) is therefore obtained from (ii) by applying the result of Problem **11(a)**.)

18. Let $\mathbf{A}$ be a finitary algebra such that $\Theta(\mathbf{A})$ is distributive. Use the results of Problems **12**, **11**, **14**, **6**(**g**) and **15** (in that sequence) to prove that the following conditions are equivalent.

(i) $\mathbf{A}$ admits a decomposition as a direct sum of simple algebras.

(ii) $\Theta(\mathbf{A}) = \Theta_0(\mathbf{A})$.

(iii) $\mathbf{A}$ has commuting congruence relations, and $\Theta(\mathbf{A})$ is a Boolean algebra.

(iv) $\mathbf{A}$ as commuting congruence relations, and every element of $\Theta(\mathbf{A})$ is a join of atoms.

The final problems of this chapter are concerned with the direct sum decompositions‡ of lattices. It is necessary to develop some of the general theory of lattice congruences in order to apply the result of Problem **18**. The general conclusions are then specialized to a particular type of lattices to obtain some basic theorems on the decomposition of lattices.

19. Throughout this problem, $\mathbf{L}$ denotes an arbitrary lattice. Recall that if $a \geqq b$, the quotient (or interval) a/b is the subset of L consisting of all x such that $a \geqq x \geqq b$. Let Q be a set of quotients of $\mathbf{L}$. Define

$$\Delta_Q = \bigcap\{\Gamma \in \Theta(\mathbf{L}) \mid \langle a, b\rangle \in \Gamma \text{ for all } a/b \in Q\}.$$

It is convenient to write $\Gamma_{a/b}$ instead of $\Gamma_{\{a/b\}}$. A number of elementary facts can be obtained immediately from this definition.

(a) Prove:

(i) if $a/b \in Q$, then $\Delta_{a/b} \subseteq \Delta_Q$;

(ii) $\langle x, y\rangle \in \Delta_Q$ if and only if $\Delta_{x\vee y/x\wedge y} \subseteq \Delta_Q$;

‡ For a lattice that has a least element o and a greatest element u, there can only be a finite number of nontrivial factors of any direct sum decomposition. Hence, in this case, the study of direct sum decompositions is equivalent to the investigation of decompositions into a direct product with finitely many factors.

(iii) $a \geqq c \geqq d \geqq b$ implies $\Delta_{c/d} \subseteq \Delta_{a/b}$;
(iv) $a = c_0 \geqq c_1 \geqq \cdots \geqq c_n = b$ implies $\Delta_{a/b} = \Delta_{c_0/c_1} \vee \cdots \vee \Delta_{c_{n-1}/c_n}$;
(v) $\Delta_{x \vee y/x} = \Delta_{y/x \wedge y}$.

These simple properties motivate the next definitions. Recall that a/b is an upper transpose of c/d (and c/d is a lower transpose of a/b) if $a/b = (x \vee y)/x$, $c/d = y/(x \wedge y)$ for some x, y in L. Equivalently, $a = b \vee c$ and $d = b \wedge c$. Let u/v and z/w be two quotients in L. Then u/v is *projective* to z/w if there is a sequence

$$u/v = a_0/b_0, a_1/b_1, \cdots, a_n/b_n = z/w$$

such that each a_{i-1}/b_{i-1} is a transpose (upper or lower) of a_i/b_i. Define u/v to be *weakly projective into* z/w if there is a sequence

$$u/v = a_0/b_0, a_1/b_1, \cdots, a_n/b_n = z/w$$

such that each a_{i-1}/b_{i-1} is contained in a transpose of a_i/b_i. A number of simple facts follow immediately from this definition and the properties listed in (a).

(b) Prove:

(i) u/v projective to z/w implies u/v weakly projective into z/w.
(ii) "Projective to" is an equivalence relation on quotients.
(iii) "Weakly projective into" is a transitive relation on quotients.
(iv) If u/v is projective to z/w, then $\Delta_{u/v} = \Delta_{z/w}$.
(v) If u/v is weakly projective into z/w, then $\Delta_{u/v} \subseteq \Delta_{z/w}$.
(vi) If $x \vee y = c_0 \geqq c_1 \geqq \cdots c_n = x \wedge y$, where each quotient c_{i-1}/c_i is weakly projective into a quotient of the set Q, then $\langle x, y\rangle \in \Delta_Q$.

Our aim is to prove the converse of this last statement. The argument depends on a simple property of weak projectivity.

(c) Let $z \geqq w$ in **L** and $u \in L$. Prove that

(i) $z \vee u/w \vee u$ is weakly projective into z/w.
(ii) $z \wedge u/w \wedge u$ is weakly projective into z/w.

(*Hint:* for the first result, show that $(z \vee u)/(w \vee u)$ is a transpose of $z/(z \wedge (w \vee u))$, which is contained in z/w. The second proof is similar.)

For any set Q of quotients in **L** it is convenient to define temporarily

$$D_Q = \{a/b \mid a = c_0 \geqq c_1 \geqq \cdots \geqq c_n = b, n \geqq 0,$$
$$\text{and each } c_{i-1}/c_i \text{ is weakly projective into some quotient of } Q\}.$$

The object is to show that $\Delta_Q = \{\langle x, y\rangle \mid (x \vee y)/(x \wedge y) \in D_Q\}$. The inclusion $\supseteq$ was noted in the last statement of **(b)**. The other inclusion will follow easily once it is established that $\{\langle x, y\rangle \mid (x \vee y)/(x \wedge y) \in D_Q\}$ is a congruence relation on **L**.

(d) Prove that $a/b \in D_Q$ implies $(a \vee c)/(b \vee c) \in D_Q$ and $(a \wedge c)/(b \wedge c) \in D_Q$.

(e) Show that if $a \geqq c \geqq d \geqq b$ and $a/b \in D_Q$, then $c/d \in D_Q$. (*Hint:* suppose that $a = e_0 > e_1 > \cdots > e_n = b$, where each e_{i-1}/e_i is weakly projective into a quotient of Q. Show that if $f_i = (e_i \vee a) \wedge b$, then $c = f_0 \geqq f_1 \geqq \cdots \geqq f_n = d$ and each f_{i-1}/f_i is weakly projective into some quotient of Q.)

(f) Prove that if $a/b \in D_Q$ and $b/c \in D_Q$, then $a/c \in D_Q$.

(g) Suppose that $(x \vee y)/(x \wedge y) \in D_Q$ and $(y \vee z)/(y \wedge z) \in D_Q$. Prove that $(x \vee z)/(x \wedge z) \in D_Q$. (*Hint:* use **(e)** and **(f)** to show that $(x \vee y \vee z)/y \in D_Q$ and $y/(z \wedge y \wedge z) \in D_Q$; then **(e)** and **(f)** give the required result.)

(h) Show that if $(x \vee y)/(x \wedge y) \in D_Q$ and $u \in L$, then $((x \vee u) \vee (y \vee u))/((x \vee u) \wedge (y \vee u)) \in D_Q$, and $((x \wedge u) \vee (y \wedge u))/((x \wedge u) \wedge (y \wedge u)) \in D_Q$.

(i) Prove: $\Delta_Q = \{\langle x, y\rangle \mid (x \vee y)/(x \wedge y) \in D_Q\}$.

To obtain a necessary and sufficient condition for a lattice to have commuting congruence relations, it is convenient to generalize the notice of "relative complement." Let $a \geqq x \geqq b$. An element $y \in a/b$ is called a *weak relative complement of x in a/b* if there is a chain $a = c_0 \geqq c_1 \geqq \cdots \geqq c_m = y = d_0 \geqq d_1 \geqq \cdots \geqq d_n = b$ such that each c_{i-1}/c_i is weakly projective into x/b and each d_{j-1}/d_j is weakly projective into a/x. A lattice **L** will be called *relatively complemented in the weak sense* if whenever $a \geqq x \geqq b$ there is a weak relative complement of x in a/b. It is obvious that if **L** is relatively complemented, then **L** is relatively complemented in the weak sense.

(j) Prove that the following conditions are equivalent for any lattice **L**:

(i) **L** has commuting congruence relations.

(ii) If $a \geqq x \geqq b$, there exists $y \in a/b$ such that $\Delta_{a/y} \subseteq \Delta_{x/b}$ and $\Delta_{y/b} \subseteq \Delta_{a/x}$.

(iii) **L** is relatively complemented in the weak sense.

(*Hints:* the result that (i) implies (ii) follows easily from the observation that if the congruences on **L** commute, then $\langle a, b\rangle \in \Delta_{x/b} \circ \Delta_{a/x}$. The proof that (ii) implies (i) follows almost verbatim the proof given in Example 1.9(b) that a relatively complemented lattice has commuting congruences. Of course, the relative complements in that argument must be replaced by weak relative complements.)

Since $\Theta(\mathbf{L})$ is a Boolean algebra if and only if its elements are joins of atoms, it is natural to try to characterize these atoms. With this end in mind, Crawley [10] has introduced the following definition: A proper quotient a/b is *minimal* if, whenever c/d is a proper quotient that is weakly projective into a/b, there exists a chain $a = e_0 \geqq e_1 \geqq \cdots \geqq e_n = b$ such that each e_{i-1}/e_i is weakly projective into c/d.

(k) Prove that the atoms of $\Theta(\mathbf{L})$ are exactly the congruences $\Delta_{a/b}$ for which a/b is a (proper) minimal quotient. (*Hints:* it is clear that every atom is of the form $\Delta_{a/b}$ for some proper quotient. If $\Delta_{a/b}$ is an atom and c/d is a proper quotient which is weakly projective into a/b, then $I_L \subseteq \Delta_{c/d} \subseteq \Delta_{a/b}$.

Since $\Delta_{a/b}$ is an atom, $\Delta_{c/d} = \Delta_{a/b}$, and the desired result follows from (**i**). The converse is similar.)

(**l**) Show that $\Theta(\mathbf{L})$ is a Boolean algebra if and only if for every proper quotient a/b of $\mathbf{L}$ there is a chain $a = c_0 > c_1 > \cdots > c_n = b$ such that c_{i-1}/c_i is minimal. (*Hint:* if $\Theta(\mathbf{L})$ is a Boolean algebra and $a > b$, then by Problem **13(c)** $\Delta_{a/b}$ is a join of atoms. Thus by (**k**), $\Delta_{a/b} = \Delta_Q$, where Q is a set of minimal quotients. Hence $a = c_0 > c_1 > \cdots > c_n = b$, where each c_{i-1}/c_i is weakly projective into a minimal quotient. From this it follows that Δ_{c_{i-1}/c_i} is an atom, and the result follows.)

Collecting these results and applying the conclusion of Problem **18**, the following structure theorem is obtained.

(**m**) Prove:

Theorem. A lattice $\mathbf{L}$ admits a decomposition as a direct sum of simple lattices if and only if

(i) $\mathbf{L}$ is relatively complemented in the weak sense, and

(ii) for every proper quotient a/b in $\mathbf{L}$ there is a finite chain $a = c_0 > c_1 > \cdots > c_n = b$ such that c_{i-1}/c_i is minimal. A lattice $\mathbf{L}$ is simple if and only if all of its proper quotients are minimal.

20. Throughout this problem, $\mathbf{L}$ denotes a relatively complemented lattice. Thus, $\mathbf{L}$ has commuting congruences. Moreover, it is possible to simplify the condition (ii) in the theorem of the preceding problem.

(**a**) Prove that if $c/d \subseteq x/(x \wedge y)$, and d_1 is a relative complement of d in $c/(x \wedge y)$, then c/d is projective to $(d_1 \vee y)/y \subseteq (x \vee y)/y$. State and prove the analogous result for the case when $c/d \subseteq (x \vee y)/y$.

(**b**) Use (**a**) to show that if c/d and a/b are intervals in a relatively complemented lattice such that c/d is weakly projective into a/b, then there exists $a_1/b_1 \subseteq a/b$ such that c/d is projective to a_1/b_1.

It is customary to say that a quotient a/b is *prime* if a covers b. A lattice is called *weakly atomic* if every proper quotient contains a prime quotient. Thus, any compactly generated lattice is weakly atomic (see Problem **4** in Chapter 2).

(**c**) Use the result of (**b**) to show that any prime quotient in a relatively complemented lattice is minimal.

(**d**) Let a/b be a proper quotient containing a prime quotient p/q. Show that a/b is minimal if and only if there is a chain $a = c_0 > c_1 > \cdots > c_n = b$ such that c_{i-1}/c_i is projective to p/q.

(**e**) Prove:

Theorem. A relatively complemented, weakly atomic lattice $\mathbf{L}$ admits a decomposition as a direct sum of simple lattices if and only if for every proper interval a/b of $\mathbf{L}$, there is a chain $a = c_0 > c_1 > \cdots > c_n = b$ such that each c_{i-1}/c_i is projective to a prime quotient. In particular, if a relatively complemented lattice satisfies one of the chain conditions (either a.c.c. or d.c.c.), then it has a decomposition as a direct sum of simple lattices. (*Hint:*

for the proof of the last statement, suppose for instance that **L** satisfies d.c.c. Let a/b be a proper quotient. From the d.c.c. it follows easily that there exists d_1 covering b. Let c_1 be a relative complement d_1 in a/b. Then a/c_1 is projective to the prime quotient d_1/b. Repeating this process yields a chain $a > c_1 > c_2 > \cdots$ which by d.c.c. must terminate in a finite number of steps at b.)

(f) Prove that if **L** is a relatively complemented lattice that satisfies either chain condition, then **L** is simple if and only if all of its prime quotients are projective. (*Hints:* if **L** is simple and p/q, r/s are prime, then $\Delta_{p/q} = \Delta_{r/s}$ and it follows easily from **(b)** and Problem **19(i)** that p/q and r/s are projective. Conversely, if all prime quotients are projective, then by **(d)**, $\Theta(\mathbf{L})$ has only one atom. Hence, **L** is simple.)

21. Let L be a modular lattice in this problem.

(a) Prove that if c/d is weakly projective into a/b, then c/d is projective to a quotient $a_1/b_1 \subseteq a/b$.

(b) Prove that prime quotients are minimal.

(c) Suppose that the quotient a/b contains a prime quotient p/q. Prove that a/b is minimal if and only if there is a chain $a = c_0 > c_1 > \cdots > c_n = b$ such that each c_{i-1}/c_i is projective to p/q.

(d) Let **L** be a weakly atomic, modular lattice. Prove that $\Theta(\mathbf{L})$ is a Boolean algebra if and only if every quotient of **L** is finite dimensional.

(e) Prove:

Theorem. Let **L** be a weakly atomic, complemented, modular lattice. Then **L** is a direct sum of simple lattices if and only if **L** is finite dimensional. In this case, the number of factors is finite (hence the direct sum is a direct product). Moreover, a finite dimensional modular lattice is simple if and only if all of its prime quotients are projective.

22. Let **L** be a distributive lattice. The first objective is to prove a standard lemma: If a/b and c/d are projective quotients in a distributive lattice, then there exists a quotient e/f such that e/f is a transpose of a/b and of c/d.

(a) Prove that if z_2/z_1 is an upper transpose of y_2/y_1, and y_2/y_1 is an upper transpose of x_2/x_1, then z_2/z_1 is an upper transpose of x_2/x_1.

(b) Let w_2/w_1 be an upper transpose of z_2/z_1, let z_2/z_1 be a lower transpose of y_2/y_1, and let y_2/y_1 be an upper transpose of x_2/x_1. Define $u_2 = y_2 \vee w_2$, $u_1 = y_1 \vee w_1$. Using the assumption that **L** is distributive, prove that u_2/u_1 is an upper transpose of both x_2/x_1 and w_2/w_1.

(c) Using **(a)**, **(b)**, and the dual of **(b)**, prove the lemma stated above by mathematical induction.

(d) Suppose that $a < b \leqq c < d$. Prove that b/a is not projective to d/c. (*Hint:* show that there cannot be a quotient e/f that is an upper transpose of both d/c and b/a. Similarly eliminate the other possibilities: e/f is an upper transpose of b/a, a lower transpose of d/c, and e/f is a lower transpose of both b/a and d/c. By the above lemma, b/a is not projective to d/c.)

(e) Let $a \geqq x \geqq b$. Prove that $A_{a/x} \cap \Delta_{x/b} = I_L$. (*Hint:* otherwise, there is a proper quotient e/f such that $\Delta_{e/f} \subseteq \Delta_{a/x}$ and $\Delta_{e/f} \subseteq \Delta_{x/b}$. By Problems **19(b)** part (i) and **21(a)** it follows that there is a proper quotient in a/x that is projective to a proper quotient in x/b. This contradicts (**d**).)

(f) Let $a \geqq x \geqq b$. Let y be a weak relative complement of x in a/b (see the definition following Problem **19(i)**). Prove that y is a relative complement of x in a/b. (*Hint:* if $x \wedge y > b$ for example, then $I_L \subset \Delta_{x \wedge y/b} \subseteq \Delta_{y/b} \cap \Delta_{x/b} \subseteq \Delta_{a/x} \cap \Delta_{x/b}$, contrary to (**e**).)

(g) Prove that if a/b is a minimal quotient in a distributive lattice, then a covers b.

(h) Prove:

Theorem. Let **L** be a distributive lattice. Then **L** admits a decomposition as a direct sum of simple lattices if and only if every proper quotient a/b of **L** is a finite Boolean algebra. A distributive lattice is simple if and only if it contains exactly two elements.

Notes to Chapter 3

Ore's theorem first appeared in the article, "On the Foundations of Abstract Algebras" [41]. This work also applies the theorem to the uniqueness problem for direct decompositions of abstract algebras. For particular kinds of algebras, other methods can be used to study direct decompositions. In particular, the Krull-Schmidt theorem gives uniqueness for the direct decompositions of certain groups with operators. The examples given in Problem **4** (due to Jónsson [28]) shows that Ore's theorem does not carry over to lattices without some finiteness condition.

Important work on direct decompositions has been done by J. Hashimoto [25]. Some of his more interesting results are sketched in Problems **10–18**. Hashimoto's paper is also the source of some of the results on the structure of lattices given in Problems **19–22**. However, it was Dilworth who initiated the study of relatively complemented lattices (see [14]).

The direct decompositions of abstract algebras and relational systems have been investigated by Jónsson and Tarski [29], Crawley and Jónsson [11], and more recently by Chang, Jónsson, and Tarski [7], using methods that are not purely lattice theoretical.

[4]

Free Algebras

1. Free Extensions

The purpose of this chapter is to introduce that part of the theory of abstract algebras that is concerned with concepts such as "free algebras" and "free sums." Both of these notions are (in a sense) special cases of the idea of a "free extension of a partial algebra." This section is devoted to the study of these free extensions; the next section deals with free sums, particularly the question of their existence; and the final section takes up free sums with amalgamation.

The methods used in this chapter are very general. They properly belong to category theory rather than to universal algebra. Most of the definitions introduced here can be reformulated entirely in terms of homomorphisms between various algebras. Consequently, they make sense* in any category. The principal results of the chapter can then be proved as theorems about suitably restricted categories. Such a program is more general than the course that we will follow. Our attention will be restricted to classes of algebras or partial algebras and the collections of all homomorphisms between them. The advantage of this more conservative policy is that it avoids the need to introduce a large assortment of new concepts. At the same time, much of the methodological flavor of category theory is preserved.

1.1 DEFINITION. Let $\mathfrak{A}$ be a class of partial algebras of type τ. Let **A** be a partial algebra of type τ. A *free* $\mathfrak{A}$*-extension* of **A** is a partial algebra **B** such that

(i) **B** is an extension of **A** and $[A] = B$.
(ii) $\mathbf{B} \in \mathfrak{A}$.
(iii) If $\mathbf{C} \in \mathfrak{A}$ and $\varphi \in \mathrm{Hom}(\mathbf{A}, \mathbf{C})$, then there exists $\psi \in \mathrm{Hom}(\mathbf{B}, \mathbf{C})$ such that $\psi \restriction A = \varphi$.

* It should be mentioned, however, that when the definitions of category theory are specialized to classes of abstract algebras, they sometimes fail to agree with the usual concepts bearing the same name. For example, in some classes of abstract algebras, the category theoretic condition for a homomorphism to be a monomorphism is satisfied by homomorphisms that are not one-to-one.

Of course, the existence of free $\mathfrak{A}$-extensions is not obvious. Indeed, the main result of this section is a sufficient condition for a partial algebra to have a free $\mathfrak{A}$-extension. Before proving this theorem, we present a simple uniqueness theorem.

1.2 PROPOSITION. Let $\mathbf{B}$ and $\mathbf{B}'$ be two free $\mathfrak{A}$-extensions of $\mathbf{A}$. Then there is an isomorphism σ of $\mathbf{B}$ to $\mathbf{B}'$, which is the identity on $\mathbf{A}$.

Proof. By Definition 1.1(iii), there exists homomorphisms $\sigma \in \text{Hom}(\mathbf{B}, \mathbf{B}')$ and $\sigma' \in \text{Hom}(\mathbf{B}', \mathbf{B})$ such that $\sigma \restriction A = \sigma' \restriction A = I_A$. Consequently, $\sigma \circ \sigma' \in \text{Hom}(\mathbf{B}', \mathbf{B}')$, $\sigma' \circ \sigma \in \text{Hom}(\mathbf{B}, \mathbf{B})$, and $(\sigma \circ \sigma') \restriction A = I_A$, $(\sigma' \circ \sigma) \restriction A = I_A$. By Lemma 1.4.12 and Definition 1.1(i), $\sigma \circ \sigma' = I_B$, and $\sigma' \circ \sigma = I_B$. By Proposition 1.2.6, σ is an isomorphism.

The basis for the proof of the fundamental existence theorem of this chapter is an estimate of the cardinality of partial algebras. This result is of interest in its own right.

1.3 PROPOSITION. Let $\mathbf{A}$ be a partial algebra of type $\tau \in (\text{Ord})^\rho$. Suppose that X is a set of generators of $\mathbf{A}$, that is, $[X] = A$. Let α be a cardinal number that satisfies the following conditions:

(i) α is infinite.
(ii) $\alpha \geqq |\rho|$.
(iii) $\alpha \geqq |X|$.
(iv) $\alpha^{|\tau(\xi)|} \leqq \alpha$ for all $\xi < \rho$.

Then $|A| \leqq \alpha$.

Proof. Let $\mathbf{A} = \langle A; F_\xi \rangle_{\xi < \rho}$. Define sets $S_\nu \subseteq A$ for $\nu < \alpha$ by the recursive conditions:

(a) $S_0 = X$.
(b) $S_\nu = \bigcup_{\zeta < \nu} S_\zeta \cup \bigcup_{\xi < \rho} F_\xi((\bigcup_{\zeta < \nu} S_\zeta)^{\tau(\xi)} \cap \mathfrak{D}(F_\xi)) \qquad \text{if } \nu > 0.$

Then by induction $|S_\nu| \leqq \alpha$ for all $\nu < \alpha$. For $\nu = 0$, this inequality is a restatement of (iii). Assuming that $|S_\nu| \leqq \alpha$ for all $\nu < \alpha$, the recursion relation (b) and the conditions (i) through (iv) yield the following estimates

$$|\bigcup_{\zeta < \nu} S_\zeta| \leqq |\nu|\, \alpha \leqq \alpha^2 = \alpha,$$
$$|F_\xi((\bigcup_{\zeta < \nu} S_\zeta)^{\tau(\xi)} \cap \mathfrak{D}(F_\xi))| \leqq |(\bigcup_{\zeta < \nu} S_\zeta)^{\tau(\xi)}| \leqq \alpha^{|\tau(\xi)|} = \alpha,$$
$$|S_\nu| \leqq \alpha + |\rho|\, \alpha = \alpha + \alpha = \alpha.$$

Thus the induction is complete. Let $S = \bigcup_{\nu < \alpha} S_\nu$. Then $|S| \leqq \alpha^2 = \alpha$. Moreover, $X \subseteq S$. Therefore the proof can be completed by showing that S is a subalgebra of $\mathbf{A}$, since in this case $A = [X] \subseteq S \subseteq A$. Suppose that $\mathbf{a} \in S^{\tau(\xi)} \cap \mathfrak{D}(F_\xi)$. Then there exists a unique mapping ν of $\tau(\xi)$ to α such that $\mathbf{a}(\eta) \in S_{\nu(\eta)}$. Let $\mu = \bigcup_{\eta < \tau(\xi)} \nu(\eta)$. Then μ is an ordinal number satisfying $\mu \leqq \alpha$ (see Propositions 5.1 and 5.2 of the Introduction). We wish to prove that $\mu < \alpha$. Suppose that $\mu = \alpha$. Then there is a mapping $\varphi = \nu \circ \psi$ of $|\tau(\xi)|$ to α (where ψ is any one-to-one mapping of $|\tau(\xi)|$ onto $\tau(\xi)$) such that $\bigcup_{\eta < |\tau(\xi)|} \varphi(\eta) = \alpha$. By Proposition 5.3(d) in the Introduction, it follows that $\alpha^{|\tau(\xi)|} > \alpha$. However, this strict inequality contradicts the condition (iv) on α. Thus $\mu < \alpha$. Consequently, $\mathbf{a} \in (\bigcup_{\zeta < \mu} S_\zeta)^{\tau(\xi)} \cap \mathfrak{D}(F_\xi)$, and by (b), $F_\xi(\mathbf{a}) \in S_{\mu+1} \subseteq S$. Hence $S \in \Sigma(\mathbf{A})$ and the proof is complete.

Remark. It is always possible to find a cardinal number α satisfying the conditions of the above proposition. For example, let

$$\alpha = (\max \{\aleph_0, |\rho|, |X|\})^\beta,$$

where $\beta = |\bigcup_{\xi < \rho} \tau(\xi)|$. Then α has the required properties.

One of the most interesting special cases of Proposition 1.3 is that in which τ is finitary and ρ is a finite or countably infinite ordinal number. Under these conditions, $\alpha = \max \{\aleph_0, |X|\}$ satisfies the conditions (i), (ii), (iii), and (iv) of 1.3. Hence, the following corollary is obtained.

1.4 COROLLARY. Let $\mathbf{A}$ be a finitary partial algebra with a finite or, at most, a countably infinite set of operations. If A is uncountable, then every set X of generators of $\mathbf{A}$ has the same cardinality as A.

It is convenient to obtain the existence of free $\mathfrak{A}$-extensions from a more general result. The extra generality will be needed in the next section.

1.5 LEMMA. Let $\mathfrak{A}$ be a class of partial algebras of type τ. Assume that† $\mathscr{S}\mathfrak{A} = \mathscr{P}\mathfrak{A} = \mathfrak{A}$. Let $\mathbf{A}$ be a partial algebra of type τ. Then there exists a partial algebra $\mathbf{B} \in \mathfrak{A}$ and a homomorphism σ of $\mathbf{A}$ to $\mathbf{B}$ such that

(a) $[\sigma(A)] = B$.

(b) If $\mathbf{C} \in \mathfrak{A}$ and $\varphi \in \mathrm{Hom}(\mathbf{A}, \mathbf{C})$, then there exists $\psi \in \mathrm{Hom}(\mathbf{B}, \mathbf{C})$ such that $\psi \circ \sigma = \varphi$.

Proof. The main step of the proof consists of showing that there is a (possibly empty) *set* of pairs

$$\{\langle A_i, \varphi_i \rangle \mid i \in I\}$$

such that

(i) $\mathbf{A}_i \in \mathfrak{A}$.

(ii) $\varphi_i \in \mathrm{Hom}(\mathbf{A}, \mathbf{A}_i)$ and $[\varphi_i(A)] = A_i$ in $\Sigma(\mathbf{A}_i)$.

(iii) If $\mathbf{C} \in \mathfrak{A}$ and $\varphi \in \mathrm{Hom}(\mathbf{A}, \mathbf{C})$, then there exists $i \in I$ and a monomorphism ϑ of $\mathbf{A}_i$ to $\mathbf{C}$ such that $\varphi = \vartheta \circ \varphi_i$.

Let α be a cardinal number that satisfies the conditions in Proposition 1.3. Fix a set X of cardinality α. Let $\{\langle \mathbf{A}_i, \varphi_i \rangle \mid i \in I\}$ be the class of all pairs $\langle \mathbf{A}_i, \varphi_i \rangle$ consisting of partial algebras $\mathbf{A}_i \in \mathfrak{A}$ with $A_i \subseteq X$ and homomorphisms φ_i of $\mathbf{A}$ to $\mathbf{A}_i$ such that $[\varphi_i(A)] = A_i$ in $\Sigma(\mathbf{A}_i)$. The class of all such

† This condition of course implies that $\mathfrak{A}$ is abstract, that is, $\mathscr{I}\mathfrak{A} = \mathfrak{A}$.

pairs is a set. In fact, each pair $\langle \mathbf{A}_i, \varphi_i \rangle$ can be identified with an element of the set $\mathbf{P}(X) \times \mathbf{X}_{\xi<\rho}\mathbf{P}(X^{\tau(\xi)+1}) \times \mathbf{P}(A \times X)$, so that there is a one-to-one correspondence between $\{\langle \mathbf{A}_i, \varphi_i \rangle \mid i \in I\}$ and a subset of a set. It is clear from the definition that $\{\langle \mathbf{A}_i, \varphi_i \rangle \mid i \in I\}$ satisfies (i) and (ii) given above. For the proof that (iii) holds, suppose that $\mathbf{C} = \langle C; H_\xi \rangle_{\xi<\rho}$ is in $\mathfrak{A}$, and φ is a homomorphism of $\mathbf{A}$ to $\mathbf{C}$. By 1.3, $|[\varphi(A)]| \leqq \alpha$. Consequently, there is a one-to-one mapping ϑ of a subset Y of X onto $[\varphi(A)]$. For $\xi < \rho$ and $\mathbf{y} \in Y^{\tau(\xi)}$, define $J_\xi(\mathbf{y}) = \vartheta^{-1}(H_\xi(\vartheta \circ \mathbf{y}))$. Then $\langle Y; J_\xi \rangle_{\xi<\rho}$ is a partial algebra of type τ and ϑ^{-1} is clearly an isomorphism of $\mathbf{C} \restriction [\varphi(A)]$ onto $\langle Y; J_\xi \rangle_{\xi<\rho}$. By Corollary 1.4.9, $[\vartheta^{-1}(\varphi(A))] = \vartheta^{-1}([\varphi(A)]) = Y$. Consequently, there is an index $i \in I$ such that $\mathbf{A}_i = \langle Y; J_\xi \rangle_{\xi<\rho}$ and $\varphi_i = \vartheta^{-1} \circ \varphi$. Thus, $\varphi = \vartheta \circ \varphi_i$, and it follows that (iii) is satisfied. To prove the lemma, let $\mathbf{D} = \Pi_{i\in I}\mathbf{A}_i$. Then $\mathbf{D} \in \mathscr{P}\mathfrak{A} = \mathfrak{A}$. By Theorem 1.5.6, there is a homomorphism σ of $\mathbf{A}$ to $\mathbf{D}$ such that $\varphi_i = \pi_i \circ \sigma$ for all $i \in I$. Define $B = [\sigma(A)]$ and $\mathbf{B} = \mathbf{D} \restriction B$. Note that $\mathbf{B} \in \mathscr{S}\mathfrak{A} = \mathfrak{A}$. If $\mathbf{C} \in \mathfrak{A}$ and $\varphi \in \operatorname{Hom}(\mathbf{A}, \mathbf{C})$, then by (iii) above there exists $i \in I$ and a monomorphism ϑ of $\mathbf{A}_i$ to $\mathbf{C}$ such that $\varphi = \vartheta \circ \varphi_i = \vartheta \circ \pi_i \circ \sigma$. Therefore, (b) is satisfied with $\psi = \vartheta \circ \pi_i$.

1.6 THEOREM. Let $\mathfrak{A}$ be a class of partial algebras of type τ. Assume that $\mathscr{S}\mathfrak{A} = \mathscr{P}\mathfrak{A} = \mathfrak{A}$. Let $\mathbf{A}$ be a partial algebra of type τ such that there is a monomorphism of $\mathbf{A}$ to some partial algebra belonging to $\mathfrak{A}$. Then there exists a free $\mathfrak{A}$-extension of $\mathbf{A}$.

Proof. Let $\mathbf{B} \in \mathfrak{A}$ and $\sigma \in \operatorname{Hom}(\mathbf{A}, \mathbf{B})$ satisfy the conditions of Lemma 1.5. Then σ is a monomorphism. In fact, by hypothesis there is a monomorphism φ of $\mathbf{A}$ to some $\mathbf{C} \in \mathfrak{A}$. Hence $\psi \in \operatorname{Hom}(\mathbf{B}, \mathbf{C})$ exists satisfying $\varphi = \psi \circ \sigma$, and therefore σ must be one-to-one. By Proposition 1.4.2, there is an extension $\mathbf{B}_0$ of $\mathbf{A}$ and an isomorphism λ of $\mathbf{B}_0$ to $\mathbf{B}$ such $\lambda \restriction A = \sigma$. Since $\mathbf{B}_0 \cong \mathbf{B}$ and $\mathbf{B} \in \mathfrak{A}$, it follows that $\mathbf{B}_0 \in \mathscr{I}\mathfrak{A} = \mathfrak{A}$. By Corollary 1.4.9, $B_0 = \lambda^{-1}(B) = \lambda^{-1}([\sigma(A)]) = [\lambda^{-1}\sigma(A)] = [A]$. Finally, suppose that $\mathbf{C} \in \mathfrak{A}$ and $\varphi \in \operatorname{Hom}(\mathbf{A}, \mathbf{C})$. By the property (b) of 1.5, there is a homomorphism ψ of $\mathbf{B}$ to $\mathbf{C}$ such that $\varphi = \psi \circ \sigma$. Let $\psi_0 = \psi \circ \lambda$. Then $\psi_0 \restriction A = \psi \circ (\lambda \restriction A) = \psi \circ \sigma = \varphi$. Therefore, $\mathbf{B}_0$ is the required free $\mathfrak{A}$-extension of A.

1.7 EXAMPLE. A partially ordered set $\mathbf{S} = \langle S; \leqq \rangle$ can be considered as a partial lattice, that is, a partial algebra with two binary partial operations $\vee$ and $\wedge$ given by

$$x \vee y = y \vee x = y \quad \text{if} \quad x \leqq y, \qquad \text{and}$$
$$x \wedge y = y \wedge x = x \quad \text{if} \quad x \leqq y.$$

Thus $x \vee y$ and $x \wedge y$ are defined if and only if x and y are comparable in the ordering of $\mathbf{S}$. It is easy to see that a mapping φ between two partially ordered sets considered as partial lattices is a homomorphism if and only if it preserves order, that is, $x \leqq y$ implies $\varphi(x) \leqq \varphi(y)$. We wish to use Theorem 1.6 in the case that $\mathfrak{A}$ is the class of all lattices. Clearly, $\mathscr{S}\mathfrak{A} = \mathscr{P}\mathfrak{A} = \mathfrak{A}$. Moreover, for any partially ordered set $\mathbf{S}$, the mapping

$x \to \{y \in S \mid y \leqq x\}$ is one-to-one order preserving mapping of S into $\mathbf{P}(S)$. Thus, $\mathbf{S}$ admits a monomorphism to a lattice. Hence, the conditions of 1.6 are fulfilled and it follows that free $\mathfrak{A}$-extensions exist for partially ordered sets. That is, every partially ordered set $\mathbf{S}$ has an extension which is a lattice $\mathbf{L}$ such that S generates $\mathbf{L}$ and every order preserving mapping of $\mathbf{S}$ into a lattice has an extension to a lattice homomorphism of $\mathbf{L}$.

An important special case of Theorem 1.6 occurs when $\mathbf{A} = \langle A; F_\xi \rangle_{\xi < \rho}$ and $F_\xi = \varnothing$ for all ξ in some subset K of ρ, while F_ξ is an operation for every $\xi \in \rho - K$. In effect, such a partial algebra is an algebra of type $\tau \restriction \rho - K$. The next example uses Theorem 1.6 in this way.

1.8 EXAMPLE. For any infinite cardinal number α, an *α-complete Boolean algebra* is defined to be a Boolean algebra $\mathbf{B}$ with the property that if $X \subseteq B$ and $|X| \leqq \alpha$ then X has a least upper bound and a greatest lower bound in $\mathbf{B}$. If $\mathbf{B}$ is an α-complete Boolean algebra, then $\mathbf{B}$ can be considered as an abstract algebra of type $\{\langle 0, 2\rangle, \langle 1, 2\rangle, \langle 2, 0\rangle, \langle 3, 0\rangle, \langle 4, 1\rangle, \langle 5, \alpha\rangle, \langle 6, \alpha\rangle\}$, with the usual Boolean algebra operations of join, meet, zero, unit, and complement, together with the α-ary operations $\bigvee \mathbf{x} = \text{l.u.b. } \{\mathbf{x}(\eta) \mid \eta < \alpha\}$ and $\bigwedge \mathbf{x} = \text{g.l.b. } \{\mathbf{x}(\eta) \mid \eta < \alpha\}$. The operations $\bigvee$ and $\bigwedge$ are called the α-join and α-meet operations in $\mathbf{B}$. Since the Boolean algebra $\mathbf{P}(X)$ of all subsets of a set X is complete, it follows from the definition that $\mathbf{P}(X)$ is an α-complete Boolean algebra. For this algebra, the α-ary operations $\bigvee$ and $\bigwedge$ are set union and set intersection.

Let $\mathfrak{A}$ be the class of all α-complete Boolean algebras. It is easily seen that $\mathscr{S}\mathfrak{A} = \mathscr{P}\mathfrak{A} = \mathfrak{A}$. By the theorem of Example 2.4.2, every Boolean algebra is isomorphic to a subalgebra of an α-complete Boolean algebra, indeed to a subalgebra of $\mathbf{P}(X)$ for some set X. Hence, Theorem 1.6 yields the following result.

1.9 THEOREM. (Yaqub [56]) Let $\mathbf{B}$ be a Boolean algebra. Then there is an extension $\mathbf{B}_\alpha$ of $\mathbf{B}$ such that $\mathbf{B}_\alpha$ is an α-complete Boolean algebra and every Boolean algebra homomorphism φ of $\mathbf{B}$ to an α-complete Boolean algebra $\mathbf{C}$ can be extended to a homomorphism ψ of $\mathbf{B}_\alpha$ to $\mathbf{C}$ that preserves the α-join and α-meet operations, that is, $\psi(\bigvee \mathbf{x}) = \bigvee \psi \circ \mathbf{x}$ and $\psi(\bigwedge \mathbf{x}) = \bigwedge \psi \circ \mathbf{x}$ for all $\mathbf{x} \in (B_\alpha)^\alpha$.

Free algebras are free extensions of partial algebras in which all of the partial operations are empty.

1.10 DEFINITION. Let $\mathfrak{A}$ be a class of similar partial algebras. A partial algebra $\mathbf{A}$ is called a *free partial $\mathfrak{A}$-algebra on the generating set X* (and X is called a *set of free generators* for $\mathbf{A}$) if

(i) $X \subseteq A$ and $[X] = A$.
(ii) $\mathbf{A} \in \mathfrak{A}$.
(iii) If $\mathbf{B} \in \mathfrak{A}$ and $\varphi \in B^X$, then there exists $\psi \in \text{Hom}(\mathbf{A}, \mathbf{B})$ such that $\psi \restriction X = \varphi$.

It is easy to see that if **A** is a free partial $\mathfrak{A}$-algebra on a set X and **B** is a free partial $\mathfrak{A}$-algebra on a set Y, and if $|X| = |Y|$, then **A** and **B** are isomorphic. This observation justifies the common terminology "**A** is the free partial $\mathfrak{A}$-algebra on α generators."

1.11 THEOREM. Let $\mathfrak{A}$ be a class of similar partial algebras. Suppose that $\mathscr{S}\mathfrak{A} = \mathscr{P}\mathfrak{A} = \mathfrak{A}$. Assume that $\mathfrak{A}$ contains some algebra with more than one element. Then for any set X there is a free partial $\mathfrak{A}$-algebra on the generating set X.

Proof. Since $\mathfrak{A}$ contains an algebra **A** with $|A| \geqq 2$ and $\mathscr{P}\mathfrak{A} = \mathfrak{A}$, it follows that there are algebras of arbitrarily large cardinality in $\mathfrak{A}$. Thus, for any set X there is a one-to-one mapping of X to some algebra of $\mathfrak{A}$. The theorem is therefore a consequence of 1.6 with X considered as a partial algebra (of the same type as the algebras in $\mathfrak{A}$) whose partial operations are all taken to be empty.

2. Free Sums

An important construction in the theory of modules is that of forming direct sums. The analogous concept in group theory is the free product. Both of these constructions are special cases of the notion of free sums of partial algebras.

2.1 DEFINITION. Let $\mathfrak{A}$ be a class of partial algebras of similarity type τ. Suppose that $\{\mathbf{A}_i \mid i \in I\}$ is a nonempty set of partial algebras of type τ. A free $\mathfrak{A}$-sum‡ of $\{\mathbf{A}_i \mid i \in I\}$ (or a free sum of $\{\mathbf{A}_i \mid i \in I\}$ in the class $\mathfrak{A}$) is a system $\langle \mathbf{B}, \sigma_i \rangle_{i \in I}$ consisting of a partial algebra $\mathbf{B} \in \mathfrak{A}$ and a set $\{\sigma_i \mid i \in I\}$ of monomorphisms $\sigma_i \in \operatorname{Hom}(\mathbf{A}_i, \mathbf{B})$ such that

(i) $[\bigcup_{i \in I} \sigma_i(A_i)] = B$ in $\Sigma(\mathbf{B})$, and

(ii) if $\mathbf{C} \in \mathfrak{A}$ and $\varphi_i \in \operatorname{Hom}(\mathbf{A}_i, \mathbf{C})$ for each $i \in I$, then there exists $\psi \in \operatorname{Hom}(\mathbf{B}, \mathbf{C})$ such that $\varphi_i = \psi \circ \sigma_i$ for all $i \in I$.

2.2 EXAMPLE. Let **R** be an associative ring. The *direct sum* of a set $\{\mathbf{M}_i \mid i \in I\}$ of **R**-modules is defined to be the submodule of $\Pi_{i \in I}\mathbf{M}_i$, which consists of all $a \in \mathsf{X}_{i \in I} M_i$ such that $\{i \in I \mid a(i) \neq 0\}$ is finite. It is usual to denote the direct sum by $\Sigma_{i \in I}\mathbf{M}_i$. For each $i \in I$, we can define a mapping of M_i to $\Sigma_{i \in I} M_i$ by $\sigma_i(a)(j) = 0$ if $j \neq i$, and $\sigma_i(a)(i) = a$. Obviously, σ_i is a monomorphism. If $a \in \Sigma_{i \in I} M_i$, and J is the finite set consisting of all i

‡ The term "free sum" is not universal by any means. However, it is consistent with the terminology "direct sum" in the theories of modules and rings. Moreover, the expression conveys the idea that free sums are dual to free (or direct) products in the sense of category theory.

such that $a(i) \neq 0$, then clearly $a = \Sigma_{j\in J}\sigma_j(a(j))$. Hence $[\bigcup_{i\in I}\sigma_i(M_i)] = \Sigma_{i\in I}M_i$. If $\mathbf{N}$ is an $\mathbf{R}$-module, and $\varphi_i \in \text{Hom}(\mathbf{M}_i, \mathbf{N})$ for all $i \in I$, define a mapping φ of $\Sigma_{i\in I}M_i$ to N as follows. Let $a \in \Sigma_{i\in I}M_i$ and $J = \{j \in I \mid a(j) \neq 0\}$. Define $\varphi(a) = \Sigma_{j\in J}\varphi_j(a(j))$. It is easy to check that φ is a homomorphism of $\Sigma_{i\in I}\mathbf{M}_i$ to $\mathbf{N}$. It is clear from the definition of φ that $\varphi_i = \varphi \circ \sigma_i$ for all $i \in I$. This discussion shows that $\langle\Sigma_{i\in I}\mathbf{M}_i; \sigma_i\rangle_{i\in I}$ is the free sum of $\{\mathbf{M}_i \mid i \in I\}$ in the class of all $\mathbf{R}$-modules.

2.3 Example. Let $\mathfrak{A}$ be a class of similar partial algebras. For each i in an index set I, suppose that $\mathbf{F}_i$ is a free partial $\mathfrak{A}$-algebra on the generating set X_i. Assume that $\langle\mathbf{F}; \sigma_i\rangle_{i\in I}$ is a free $\mathfrak{A}$-sum of $\{\mathbf{F}_i \mid i \in I\}$. Define $X = \bigcup_{i\in I}\sigma_i(X_i)$. We will prove that $\mathbf{F}$ is a free partial $\mathfrak{A}$-algebra on the generating set X. First note that $[X] = [\bigcup_{i\in I}\sigma_i(X_i)] = [\bigcup_{i\in I}[\sigma_i(X_i)]] \supseteq [\bigcup_{i\in I}\sigma_i([X_i])] = [\bigcup_{i\in I}\sigma_i(F_i)] = F$. Therefore, X generates $\mathbf{F}$. If φ is any mapping of X to $\mathbf{B} \in \mathfrak{A}$, then $\varphi \circ (\sigma_i \upharpoonright X_i)$ can be extended to a homomorphism ψ_i of $\mathbf{F}_i$ to $\mathbf{B}$. Then there is a homomorphism ψ of $\mathbf{F}$ to $\mathbf{B}$ such that $\psi \circ \sigma_i = \psi_i$ for all $i \in I$. If $x_i \in X_i$, then $(\psi \circ \sigma_i)(x_i) = \psi_i(x_i) = (\varphi \circ \sigma_i)(x_i)$. Therefore, $\psi \upharpoonright X = \varphi$. This proves the claim that $\mathbf{F}$ is a free partial $\mathfrak{A}$-algebra on the generating set X.

Just as in the case of free $\mathfrak{A}$-extensions, the question of uniqueness is easily settled. Only the existence problem poses some difficulty.

2.4 Proposition. Let $\langle\mathbf{B}; \sigma_i\rangle_{i\in I}$ and $\langle\mathbf{B}'; \sigma'\rangle_{i\in I}$ be two free $\mathfrak{A}$-sums of $\{\mathbf{A}_i \mid i \in I\}$. Then there is an isomorphism σ of B' to B such that $\sigma_i = \sigma \circ \sigma'_i$ for all $i \in I$.

Proof. By condition 2.1(ii) applied to both of the free sums, there exist homomorphisms σ of $\mathbf{B}'$ to $\mathbf{B}$ and σ' of $\mathbf{B}$ to $\mathbf{B}'$ such that $\sigma_i = \sigma \circ \sigma'_i$ and $\sigma'_i = \sigma' \circ \sigma_i$ for all $i \in I$. Therefore, $\sigma_i = \sigma \circ \sigma' \circ \sigma_i$ for all $i \in I$. It follows that $\sigma \circ \sigma'$ is the identity mapping on $\bigcup_{i\in I}\sigma_i(A_i)$. Consequently, by Lemma 1.4.12 and condition (i) of Definition 2.1, $\sigma \circ \sigma' = I_B$. Similarly $\sigma' \circ \sigma = I_{B'}$. Hence σ is an isomorphism by Proposition 1.2.6.

The main result of this section is an existence theorem for free sums of algebras. It is possible to prove a more general result that is applicable to partial algebras. However, the restriction to algebras allows us to eliminate many uninteresting technicalities from the proof and the result for algebras is general enough for most applications. A sketch of the way in which the more general result is obtained will be found in Problem **3**.

2.5 Theorem. Let $\mathfrak{A}$ be a class of abstract algebras of type τ. Suppose that $\mathscr{S}\mathfrak{A} = \mathscr{P}\mathfrak{A} = \mathfrak{A}$. Let $\{\mathbf{A}_i \mid i \in I\}$ be a nonempty subset of $\mathfrak{A}$, which has the following *embedding property* (EP). There exists $\mathbf{C}_0 \in \mathfrak{A}$ and a set $\{\psi_i \mid i \in I\}$ of monomorphisms with $\psi_i \in \text{Hom}(\mathbf{A}_i, \mathbf{C}_0)$. Then the free $\mathfrak{A}$-sum of $\{\mathbf{A}_i \mid i \in I\}$ exists.

There are two steps in the proof of this theorem. First, we show by direct construction that under the hypothesis of the theorem, the free sum of

$\{\mathbf{A}_i \mid i \in I\}$ exists in the class of all partial algebras of type τ. Roughly speaking, the free sum in this class of partial algebras is obtained by taking the disjoint union of copies of the $\mathbf{A}_i$ and identifying certain distinguished elements. The second part of the proof consists of showing that Lemma 1.5 can be applied to this free sum to yield the desired free $\mathfrak{A}$-sum.

It is convenient to introduce some notation and to prove a preliminary fact.

2.6 Definition. Let $\mathbf{A}$ be a partial algebra. Define $K(\mathbf{A}) = [\varnothing]$ and $\mathbf{K}(\mathbf{A}) = \mathbf{A} \upharpoonright K(\mathbf{A})$. That is, $\mathbf{K}(\mathbf{A})$ is the smallest subalgebra of $\mathbf{A}$. The partial algebra $\mathbf{K}(\mathbf{A})$ will be called the *prime subalgebra* of $\mathbf{A}$, and the elements of $K(\mathbf{A})$ will be called the *constants* of $\mathbf{A}$.

Note that $K(\mathbf{A})$ contains all elements of the form $F_\xi(0)$ where F_ξ is a zero-ary operation of $\mathbf{A}$. Moreover, if $\mathbf{A}$ has no zero-ary operations, then $K(\mathbf{A}) = \varnothing$.

2.7 Lemma. Let $\mathbf{A}$ and $\mathbf{B}$ be algebras of type τ. Suppose that $\varphi \in \operatorname{Hom}(\mathbf{A}, \mathbf{B})$ is a monomorphism. Then $\varphi \upharpoonright K(\mathbf{A})$ is an isomorphism of $\mathbf{K}(\mathbf{A})$ to $\mathbf{K}(\mathbf{B})$. Moreover, if $\psi_0 \in \operatorname{Hom}(\mathbf{A}, \mathbf{B})$ and $\psi_1 \in \operatorname{Hom}(\mathbf{A}, \mathbf{B})$, then $\psi_0 \upharpoonright K(\mathbf{A}) = \psi_1 \upharpoonright K(\mathbf{A})$.

Proof. By 1.4.11, $\varphi(K(\mathbf{A})) = \varphi([\varnothing]) = [\varphi(\varnothing)] = [\varnothing] = K(\mathbf{B})$. Thus by Proposition 1.2.4, φ is an epimorphism of $\boldsymbol{K}(\mathbf{A})$ to $\boldsymbol{K}(\mathbf{B})$. Since φ is a monomorphism by assumption, the first statement of the lemma is proved. The last statement follows from the case of Lemma 1.4.12 in which $X = \varnothing$.

Proof of Theorem 2.5. To establish notation, let $\mathbf{A}_i = \langle A_i; F_{\xi i}\rangle_{\xi<\rho}$ for each $i \in I$, and let $\mathbf{C}_0 = \langle C_0; H_\xi\rangle_{\xi<\rho}$. Note that by Lemma 2.7, $\psi_i \upharpoonright K(\mathbf{A}_i)$ is an isomorphism of $\boldsymbol{K}(\mathbf{A}_i)$ to $\boldsymbol{K}(\mathbf{C}_0)$.

(1) We construct the free sum of $\{\mathbf{A}_i \mid i \in I\}$ in the class of all partial algebras of type τ as follows. Let $\{X_i \mid i \in I\}$ be a family of sets satisfying

(a) $X_i \cap K(\mathbf{C}_0) = \varnothing$ for all i.
(b) $X_i \cap X_j = \varnothing$ if $i \neq j$.
(c) $|X_i| = |A_i - K(\mathbf{A}_i)|$.

Let λ_i be a one-to-one mapping of A_i onto $K(\mathbf{C}_0) \cup X_i$ such that

(d) $\lambda_i \upharpoonright K(\mathbf{A}_i) = \psi_i \upharpoonright K(\mathbf{A}_i)$.

By (a), (c), and Lemma 2.7, such a λ_i exists. Define

(e) $A = K(\mathbf{C}_0) \cup \bigcup_{i\in I} X_i = \bigcup_{i\in I}\lambda_i(A_i)$.
(f) $F_\xi = \bigcup_{i\in I}\lambda_i \circ F_{\xi i}$.

It is necessary to show that F_ξ is a partial operation on A. Clearly, F_ξ is a $(\tau(\xi)+1)$-ary relation on A. Suppose that $\mathbf{a} \in \mathscr{D}(F_{\xi i})$ and $\mathbf{b} \in \mathscr{D}(F_{\xi j})$ are such that $\lambda_i \circ \mathbf{a} = \lambda_j \circ \mathbf{b}$. If $i = j$, then $\mathbf{a} = \mathbf{b}$ because λ_i is one-to-one. In this case, $\lambda_i(F_{\xi i}(\mathbf{a})) = \lambda_j(F_{\xi j}(\mathbf{b}))$ clearly. Suppose that $i \neq j$. From the equality $\lambda_i \circ \mathbf{a} = \lambda_j \circ \mathbf{b}$ and the condition (b) it follows that $\lambda_i \circ \mathbf{a}$ (and of course $\lambda_j \circ \mathbf{b}$) is in $K(\mathbf{C}_0)^{\tau(\xi)}$. Thus, $\mathbf{a} \in K(\mathbf{A}_i)^{\tau(\xi)}$ and $\mathbf{b} \in K(\mathbf{A}_j)^{\tau(\xi)}$ and $\psi_i \circ \mathbf{a} = \lambda_i \circ \mathbf{a} = \lambda_j \circ \mathbf{b} = \psi_j \circ \mathbf{b}$, so that, $F_{\xi i}(\mathbf{a}) \in K(\mathbf{A}_i)$, $F_{\xi j}(\mathbf{b}) \in K(\mathbf{A}_j)$, and $\lambda_i(F_{\xi i}(\mathbf{a})) = \psi_i(F_{\xi i}(\mathbf{a})) = H_\xi(\psi_i \circ \mathbf{a}) = H_\xi(\psi_j \circ \mathbf{b}) = \psi_j(F_{\xi j}(\mathbf{b})) = \lambda_j(F_{\xi j}(\mathbf{b}))$. This proves that F_ξ is a partial operation on A. Therefore $\mathbf{A} = \langle A; F_\xi \rangle_{\xi < \rho}$ is a partial algebra of type τ. Moreover, by the definition of F_ξ, λ_i is a monomorphism of $\mathbf{A}_i$ to $\mathbf{A}$. We wish to prove that $\langle \mathbf{A}; \lambda_i \rangle_{i \in I}$ is a free sum of $\{\mathbf{A}_i \mid i \in I\}$ in the class of all partial algebras of type τ. Because of (e), the condition (i) of Definition 2.1 is certainly satisfied. Suppose that $\mathbf{C} = \langle C; J_\xi \rangle_{\xi < \rho}$ is a partial algebra of type τ, and $\{\varphi_i \mid i \in I\}$ is a set of homomorphisms with $\varphi_i \in \text{Hom}(\mathbf{A}_i, \mathbf{C})$. Let $\varphi = \bigcup_{i \in I} \varphi_i \circ \lambda_i^{-1}$. It will be shown first that φ is a mapping. Let $x \in A = K(\mathbf{C}_0) \cup \bigcup_{i \in I} X_i$. If $x \in X_i$ for some i, then by (a) and (b), $x \in \mathscr{D}(\lambda_j^{-1})$ if and only if $j = i$. Therefore, since λ_i is one-to-one, $\varphi(x)$ is well defined. Suppose that $x \in K(\mathbf{C}_0)$. By Lemma 2.7, $x \in \mathscr{D}(\varphi_i \circ \psi_i^{-1}) \subseteq \mathscr{D}(\varphi_i \circ \lambda_i^{-1})$ for all i, and $(\varphi_i \circ \lambda_i^{-1})(x) = \varphi_i(\psi_i^{-1}(x)) = \varphi_j(\lambda_j^{-1}(x)) = (\varphi_j \circ \lambda_j^{-1})(x)$ for all i and j. Therefore, $\varphi(x)$ is well defined in this case also. By (f) and what has just been noted, $\varphi \circ F_\xi = \varphi \circ (\bigcup_{i \in I} \lambda_i \circ F_{\xi i}) = \bigcup_{i \in I} \varphi_i \circ \lambda_i^{-1} \circ \lambda_i \circ F_{\xi i} = \bigcup_{i \in I} \varphi_i \circ F_{\xi i} \subseteq J_\xi$. Therefore, $\varphi \in \text{Hom}(\mathbf{A}, \mathbf{C})$. Finally,

$$\varphi \circ \lambda_i = \bigcup_{j \in I} (\varphi_j \circ \lambda_j^{-1}) \circ \lambda_i = \varphi_i \cup \bigcup_{j \neq i} \varphi_j (\lambda_j^{-1} \circ \lambda_i).$$

By (a) and (b), $\lambda_j^{-1} \circ \lambda_i = (\lambda_j^{-1} \circ \lambda_i) \restriction K(\mathbf{A}_i)$. Thus, $\varphi_j \circ (\lambda_j^{-1} \circ \lambda_i)$ is a homomorphism from $K(\mathbf{A}_i)$ to $K(\mathbf{C}_0)$. It follows from Lemma 2.7 that $\varphi_j \circ (\lambda_j^{-1} \circ \lambda_i) = (\varphi_i \circ (\lambda_i^{-1} \circ \lambda_i)) \restriction K(\mathbf{A}_i) = \varphi_i \restriction K(\mathbf{A}_i)$: (Note that $K(\mathbf{K}(\mathbf{A}_i)) = K(\mathbf{A}_i)$ and $K(\mathbf{K}(\mathbf{C}_0)) = K(\mathbf{C}_0)$.) Therefore, $\varphi \circ \lambda_i = \varphi_i$. This shows that $\langle \mathbf{A}; \lambda_i \rangle_{i \in I}$ satisfies condition (ii) of Definition 2.1 so that $\langle \mathbf{A}, \lambda_i \rangle_{i \in I}$ is a free sum of $\{\mathbf{A}_i \mid i \in I\}$ in the class of all partial algebras of type τ.

(2) We now apply Lemma 1.5 to the partial algebra $\mathbf{A}$. Thus, there is an algebra $\mathbf{B} \in \mathfrak{A}$ and a homomorphism σ of $\mathbf{A}$ to $\mathbf{B}$ satisfying the conditions 1.5(a) and (b). Define $\sigma_i = \sigma \circ \lambda_i$. The proof of 2.5 is completed by showing that $\langle \mathbf{B}; \sigma_i \rangle_{i \in I}$ is a free $\mathfrak{A}$-sum of $\{\mathbf{A}_i \mid i \in I\}$. Property (i) of Definition 2.1 is obtained using (e) and the property (a) of Lemma 1.5: $[\bigcup_{i \in I} \sigma_i(A_i)] = [\sigma(\bigcup_{i \in I} \lambda_i(A_i))] = [\sigma(A)] = B$. To obtain (ii) of Definition 2.1, suppose that $\mathbf{C} \in \mathfrak{A}$, and $\varphi_i \in \text{Hom}(\mathbf{A}_i, \mathbf{C})$ for each $i \in I$. Since $\langle \mathbf{A}; \lambda_i \rangle_{i \in I}$ is a free sum of $\{\mathbf{A}_i \mid i \in I\}$ in the class of all partial algebras of type τ, there exists $\varphi \in \text{Hom}(\mathbf{A}, \mathbf{C})$ such that $\varphi \circ \lambda_i = \varphi_i$ for all $i \in I$. By (6) of Lemma 1.5, there exists $\psi \in \text{Hom}(\mathbf{B}, \mathbf{C})$ such that $\psi \circ \sigma = \varphi$. Therefore, $\psi \circ \sigma_i = \psi \circ \sigma \circ \lambda_i = \varphi \circ \lambda_i = \varphi_i$ for all $i \in I$. Finally, we note that the homomorphisms σ_i must be one-to-one. Indeed, by what has just been shown, there is a homomorphism ψ of $\mathbf{B}$ to $\mathbf{C}_0$ such that $\psi \circ \sigma_i = \psi_i$ (where $\mathbf{C}_0$ and the ψ_i satisfy the condition (EP)). Since all ψ_i are monomorphisms, it follows that the σ_i are also one-to-one.

In using Theorem 2.5, it is often difficult to show that the condition (EP) is satisfied. This condition can be replaced by a somewhat weaker statement.

2.8 COROLLARY. Let $\mathfrak{A}$ be a class of similar algebras such that $\mathscr{S}\mathfrak{A} = \mathscr{P}\mathfrak{A} = \mathfrak{A}$. Let $\{\mathbf{A}_i \mid i \in I\}$ be a subset of $\mathfrak{A}$ such that each $\mathbf{A}_i$ is a subalgebra of some $\mathbf{B}_i$ in $\mathfrak{A}$ with the property that Hom $(\mathbf{A}_i, \mathbf{B}_j) \neq \varnothing$ for all i and j. Then $\{\mathbf{A}_i \mid i \in I\}$ has a free $\mathfrak{A}$-sum.

Proof. For each pair $\langle i, j\rangle$, choose $\psi_{i,j} \in$ Hom $(\mathbf{A}_i, \mathbf{B}_j)$, with $\psi_{i,i} = I_{A_i}$. Let $\mathbf{C}_0 = \Pi_{j \in I}\mathbf{B}_j$. Then $\mathbf{C}_0 \in \mathscr{P}\mathfrak{A} = \mathfrak{A}$. By Theorem 1.5.6, there is a homomorphism ψ_i of $\mathbf{A}_i$ to $\mathbf{C}_0$ such that $\psi_{i,j} = \pi_j \circ \psi_i$ for all $j \in I$. Since $\psi_{i,i}$ is a monomorphism, so is ψ_i. Hence the condition (EP) is satisfied.

A particular application of Corollary 2.8 occurs when all of the algebras $\mathbf{A}_i$ have a one-element subalgebra. In this case, the mapping that sends A_i onto the one-element subalgebra of $\mathbf{A}_j$ is a homomorphism, so that Hom $(\mathbf{A}_i, \mathbf{A}_j) \neq \varnothing$. Thus, Corollary 2.8 can be used with $\mathbf{B}_i = \mathbf{A}_i$ to obtain the existence of free $\mathfrak{A}$-sums for many classes of algebras. For example, if $\mathfrak{A}$ is any one of the following classes, then every subset of $\mathfrak{A}$ has a free sum:

$\mathfrak{A}$ = all groups.
$\mathfrak{A}$ = all modules over a ring $\mathbf{R}$.
$\mathfrak{A}$ = all lattices.
$\mathfrak{A}$ = all rings.

3. Injective Completeness

It follows from Lemma 2.7 that if $\mathfrak{A}$ is the class of all rings with an identity (that is, the identity is considered as a constant of the ring), then not all subsets of $\mathfrak{A}$ have a free sum. The difficulty is that if two rings have a free sum, then the subrings generated by the identities of these rings must be isomorphic. In general, for a set $\{\mathbf{A}_i \mid i \in I\}$ of algebras to admit a free $\mathfrak{A}$-sum it is necessary that the subalgebras $\mathbf{K}(\mathbf{A}_i)$ be isomorphic. Usually, this condition is not sufficient. However, if strong enough restrictions are imposed on $\mathfrak{A}$, the isomorphism of the prime subalgebras becomes a sufficient condition for the existence of the free sum.

3.1 DEFINITION. Let $\mathfrak{A}$ be a class of partial algebras of type τ. A partial algebra $\mathbf{D}$ (not necessarily in $\mathfrak{A}$) is called *injective for* $\mathfrak{A}$ if the conditions:

(i) $\mathbf{A}$ and $\mathbf{B}$ are algebras in $\mathfrak{A}$,
(ii) $\vartheta \in$ Hom $(\mathbf{A}, \mathbf{B})$ and $\varphi \in$ Hom $(\mathbf{A}, \mathbf{D})$, and
(iii) ϑ is a monomorphism

imply that there is a homomorphism ψ of $\mathbf{B}$ to $\mathbf{D}$ such that $\varphi = \psi \circ \vartheta$. In case $\mathbf{D} \in \mathfrak{A}$, we will say that $\mathbf{D}$ is *injective in* $\mathfrak{A}$. A class $\mathfrak{A}$ of similar partial algebras is said to be *injectively complete* if every $\mathbf{A} \in \mathfrak{A}$ has an extension which is injective in $\mathfrak{A}$.

Injective completeness is a strong condition for a class of algebras, and to prove that some particular class of algebras has this property usually requires a nontrivial argument. It is well known that the class of all modules over an arbitrary (fixed) ring **R** is injectively complete (see [6], p. 6, or p. 132). We will presently prove that the class of all Boolean algebras is injectively complete. Before doing so however, it seems desirable to see why injective completeness is important in the study of free sums.

3.2 COROLLARY. Let $\mathfrak{A}$ be an injectively complete class of similar algebras such that $\mathscr{S}\mathfrak{A} = \mathscr{P}\mathfrak{A} = \mathfrak{A}$. Let $\{\mathbf{A}_i \mid i \in I\}$ be a subset of $\mathfrak{A}$. Then the free $\mathfrak{A}$-sum of $\{\mathbf{A}_i \mid i \in I\}$ exists if and only if $\mathbf{K}(\mathbf{A}_i) \cong \mathbf{K}(\mathbf{A}_j)$ for all i and j in I.

Proof. If $\langle \mathbf{A}; \sigma_i \rangle_{i \in I}$ is a free $\mathfrak{A}$-sum of $\{\mathbf{A}_i \mid i \in I\}$, then $\mathbf{K}(\mathbf{A}) \cong \mathbf{K}(\mathbf{A}_i)$ by Lemma 2.7. Hence, $\mathbf{K}(\mathbf{A}_i) \cong \mathbf{K}(\mathbf{A}_j)$ for all i and j. Conversely, suppose that for each i and j there is an isomorphism φ_{ij} of $\mathbf{K}(\mathbf{A}_j)$ to $\mathbf{K}(\mathbf{A}_i)$. Let $\mathbf{D}_i$ be an extension of $\mathbf{A}_i$ that is injective in $\mathfrak{A}$. Such an algebra exists by the injective completeness of $\mathfrak{A}$. Then $\varphi_{ij} \in \mathrm{Hom}\,(\mathbf{K}(\mathbf{A}_j), \mathbf{D}_i)$. Since $\mathbf{D}_i$ is injective, there is a homomorphism ψ_{ij} of $\mathbf{A}_j$ to $\mathbf{D}_i$ which extends φ_{ij}. Therefore, the free $\mathfrak{A}$-sum of $\{\mathbf{A}_i \mid i \in I\}$ exists by Corollary 2.8.

The concept of a free product with an amalgamated subgroup plays an important part in the theory of infinite groups. This notion can easily be generalized to abstract algebras. For consistency, we will call these objects free sums with amalgamated subalgebras.

Let $\mathfrak{A}$ be a class of algebras of type τ. Let $\mathbf{E}$ be an algebra of type τ (not necessarily in $\mathfrak{A}$). Suppose that $\{\mathbf{A}_i \mid i \in I\}$ is a subset of $\mathfrak{A}$, and that for each $i \in I$, λ_i is a monomorphism of $\mathbf{E}$ to $\mathbf{A}_i$. A free sum of $\{\mathbf{A}_i \mid i \in I\}$ in $\mathfrak{A}$ with amalgamated subalgebra $\mathbf{E}$ is a system $\langle \mathbf{B}; \sigma_i \rangle_{i \in I}$ such that $\mathbf{B}$ is in $\mathfrak{A}$, each σ_i is a monomorphism of $\mathbf{A}_i$ to $\mathbf{B}$, and the following conditions are satisfied:

(i) $\sigma_i \circ \lambda_i = \sigma_j \circ \lambda_j$ for all i and j in I.

(ii) $[\bigcup_{i \in I} \sigma_i(A_i)] = B$ in $\Sigma(\mathbf{B})$.

(iii) If $\mathbf{C} \in \mathfrak{A}$ and for each $i \in I$, $\varphi_i \in \mathrm{Hom}\,(\mathbf{A}_i, \mathbf{C})$ is given so that $\varphi_i \circ \lambda_i = \varphi_j \circ \lambda_j$ for all i and j in I, then there exists $\varphi \in \mathrm{Hom}\,(\mathbf{B}, \mathbf{C})$ such that $\varphi_i = \varphi \circ \sigma_i$ for all $i \in I$.

For our purposes, it is convenient to reformulate this definition, so that the results of section 2 can be used directly to prove the existence of free sums with amalgamation.

3.3 DEFINITION. Let $\mathfrak{A}$ be a class of algebras of type $\tau \in (\mathrm{Ord})^\rho$. Let $\mathbf{E}$ be an algebra of type τ such that $E \cap \rho = \varnothing$. Define

$$\tau_E = \tau \cup \{\langle e, 0 \rangle \mid e \in E\} \in (\mathrm{Ord})^{\rho \cup E}.$$

The $\mathbf{E}$-*amalgamation class* of $\mathfrak{A}$ (denoted by $\mathfrak{A}(\mathbf{E})$) is the class of all algebras

$\mathbf{A} = \langle A; F_\alpha \rangle_{\alpha \in \rho \cup E}$ of type τ_E such that

(i) $\langle A; F_\xi \rangle_{\xi < \rho}$ is in $\mathfrak{A}$, and

(ii) the mapping that sends an element e of E into $F_e(0)$ is a monomorphism of $\mathbf{E}$ to $\langle A; F_\xi \rangle_{\xi < \rho}$.

If $\{\mathbf{A}_i \mid i \in I\}$ is a subset of $\mathfrak{A}(\mathbf{E})$, then a free $\mathfrak{A}(\mathbf{E})$-sum of this set is called a *free* $\mathfrak{A}$*-sum of* $\{\mathbf{A}_i \mid i \in I\}$ *with amalgamated subalgebra* $\mathbf{E}$.

An element of $\mathfrak{A}(\mathbf{E})$ can be considered as an algebra $\mathbf{A}$ of $\mathfrak{A}$, together with a distinguished monomorphism of $\mathbf{E}$ to $\mathbf{A}$. Indeed, any monomorphism λ of $\mathbf{E}$ to $\mathbf{A} \in \mathfrak{A}$ induces additional zero-ary operations $F_e(0) = \lambda(e)$, corresponding to each $e \in E$, which make $\mathbf{A}$ into an algebra of $\mathfrak{A}(\mathbf{E})$. This observation is the basis of the equivalence between 3.3 and the definition that preceded it.

It is evident from Proposition 1.4.10 that if $\mathbf{A} \in \mathfrak{A}(\mathbf{E})$, then $K(\mathbf{A}) = \{F_e(0) \mid e \in E\}$, which determines a subalgebra of $\langle A; F_\xi \rangle_{\xi < \rho}$ isomorphic to $\mathbf{E}$. This remark, together with Corollary 3.2 justifies the last statement of the following result.

3.4 THEOREM. Let $\mathfrak{A}$ be an injectively complete class of algebras of type τ such that $\mathscr{S}\mathfrak{A} = \mathscr{P}\mathfrak{A} = \mathfrak{A}$. Let $\mathbf{E} \in \mathfrak{A}$. Then $\mathfrak{A}(\mathbf{E})$ is injectively complete, and $\mathscr{S}\mathfrak{A}(\mathbf{E}) = \mathscr{P}\mathfrak{A}(\mathbf{E}) = \mathfrak{A}(\mathbf{E})$. Hence every subset of $\mathfrak{A}(\mathbf{E})$ has a free $\mathfrak{A}$-sum with amalgamated subalgebra $\mathbf{E}$.

Proof. It is obvious that $\mathscr{S}\mathfrak{A}(\mathbf{E}) = \mathscr{P}\mathfrak{A}(\mathbf{E}) = \mathfrak{A}(\mathbf{E})$. Let $\mathbf{A} = \langle A; F_\alpha \rangle_{\alpha \in \rho \cup E}$ be an algebra belonging to $\mathfrak{A}(\mathbf{E})$. Then there exists an extension $\mathbf{D} = \langle D; J_\xi \rangle_{\xi < \rho}$ (in $\mathfrak{A}$) of $\langle A; F_\xi \rangle_{\xi < \rho}$, such that $\mathbf{D}$ is injective in $\mathfrak{A}$. Since $\{F_e(0) \mid e \in E\} \subseteq A \subseteq D$, it is possible to define $\mathbf{D}' = \langle D; J_\xi, F_e \rangle_{\xi < \rho, e \in E}$. Clearly, $\mathbf{D}' \in \mathfrak{A}(\mathbf{E})$. The proof of the theorem will be completed by showing that $\mathbf{D}'$ is injective in $\mathfrak{A}(\mathbf{E})$. Let $\mathbf{B} = \langle B; G_\alpha \rangle_{\alpha \in \rho \cup E}$ and $\mathbf{C} = \langle C; H_\alpha \rangle_{\alpha \in \rho \cup E}$ be in $\mathfrak{A}(\mathbf{E})$, and $\varphi \in \mathrm{Hom}\,(\mathbf{B}, \mathbf{D}')$, $\vartheta \in \mathrm{Hom}\,(\mathbf{B}, \mathbf{C})$, where ϑ is a monomorphism. Since $\mathbf{D}$ is injective and $\langle B; G_\xi \rangle_{\xi < \rho}$ and $\langle C; H_\xi \rangle_{\xi < \rho}$ are in $\mathfrak{A}$, it follows that there is a homomorphism ψ of $\langle C; H_\xi \rangle_{\xi < \rho}$ to $\mathbf{D}$ such that $\varphi = \psi \circ \vartheta$. It remains to note that $\psi \in \mathrm{Hom}\,(\mathbf{C}, \mathbf{D})$, that is, $\vartheta(H_e(0)) = F_e(0)$. In fact, $\vartheta(H_e(0)) = \vartheta(\psi(G_e(0))) = \varphi(G_e(0)) = F_e(0)$.

3.5 EXAMPLE. We will prove that the class $\mathfrak{B}$ of all Boolean algebras is injectively complete. This class obviously satisfies $\mathscr{S}\mathfrak{B} = \mathscr{P}\mathfrak{B} = \mathfrak{B}$. Thus, Theorem 3.4 implies that free sums with amalgamation exist for the class of all Boolean algebras (a result which is due to Dwinger and Yaqub [17]). The proof that $\mathfrak{B}$ is injectively complete is based on a general lemma that is valid for an arbitrary class $\mathfrak{A}$ of similar partial algebras satisfying $\mathscr{P}\mathfrak{A} = \mathfrak{A}$.

3.6 LEMMA. Let $\{\mathbf{D}_i \mid i \in I\}$ be a subset§ of $\mathfrak{A}$, consisting of partial algebras that are injective in $\mathfrak{A}$. Then $\Pi_{i \in I} \mathbf{D}_i$ is injective in $\mathfrak{A}$.

§ Note that this result is even valid when $I = \varnothing$. In this case the direct product is the one element algebra, which is clearly injective in $\mathfrak{A}$.

Proof. Let **B**, **C** be in $\mathfrak{A}$, and $\varphi \in \text{Hom}(\mathbf{B}, \Pi_{i \in I}\mathbf{D}_i)$, $\vartheta \in \text{Hom}(\mathbf{B}, \mathbf{C})$, where ϑ is a monomorphism. Then $\pi_i \circ \varphi \in \text{Hom}(\mathbf{B}, \mathbf{D}_i)$, and since $\mathbf{D}_i$ is injective, there exists $\psi_i \in \text{Hom}(\mathbf{C}, \mathbf{D}_i)$ such that $\pi_i \circ \varphi = \psi_i \circ \vartheta$. By Theorem 1.5.6, there is a homomorphism ψ of **C** to $\Pi_{i \in I}\mathbf{D}_i$ such that $\psi_i = \pi_i \circ \psi$ for all $i \in I$. Thus, $\pi_i \circ \varphi = \pi_i \circ \psi \circ \vartheta$ for all i. By 1.5.2, $\varphi = \psi \circ \vartheta$. This shows that $\Pi_{i \in I}\mathbf{D}_i$ is injective in $\mathfrak{A}$.

Next, we prove a fact about Boolean algebras.

3.7 LEMMA. Let **A** and **B** be Boolean algebras, with **A** a subalgebra of **B**. Suppose that $\Delta \in \Theta(\mathbf{A})$. Then there exists $\Gamma \in \Theta(\mathbf{B})$ such that $\Delta = \Gamma \cap A^2$.

Proof. Let $J = \bar{\Delta}^{-1}(0) = \{a \in A \mid \langle a, 0\rangle \in \Delta\}$. Note that if $a \in J$ and $b \in J$, then $a \vee b \in J$. Define Γ to be the set of all pairs $\langle x, y\rangle$ in B^2 such that $x \vee a = y \vee a$ for some $a \in J$. It is obvious that the conditions (i) and (ii) in the Definition 1.3.1 of an equivalence relation are satisfied. To prove that the condition (iii) holds, suppose that $\langle x, y\rangle \in \Gamma$ and $\langle y, z\rangle \in \Gamma$. Then a and b exist in J such that $x \vee a = y \vee a$, $y \vee b = z \vee b$. Consequently, $x \vee (a \vee b) = (x \vee a) \vee b = (y \vee a) \vee b = (y \vee b) \vee a = (z \vee b) \vee a = z \vee (a \vee b)$. Since $a \vee b \in J$, it follows that $\langle x, z\rangle \in \Gamma$. To prove that Γ is a congruence relation, suppose that $\langle x, y\rangle \in \Gamma$ and $\langle z, w\rangle \in \Gamma$. Then $x \vee a = y \vee a$ and $z \vee b = w \vee b$ for some a and b in J, so that $(x \vee z) \vee (a \vee b) = (x \vee a) \vee (z \vee b) = (y \vee a) \vee (w \vee b) = (y \vee w) \vee (a \vee b)$ and $(x \wedge z) \vee (a \vee b) = (x \vee a \vee b) \wedge (z \vee b \vee a) = (y \vee a \vee b) \wedge (w \vee b \vee a) = (y \wedge w) \vee (a \vee b)$. Hence, $\langle x \vee z, y \vee w\rangle \in \Gamma$ and $\langle x \wedge z, y \wedge w\rangle \in \Gamma$. The fact that $x \vee a = y \vee a$ implies $x' \vee a = y' \vee a$ was proved in the course of Example 2.4.2. Hence $\langle x', y'\rangle \in \Gamma$. Therefore, Γ is a congruence relation on **B**. If $\langle b, c\rangle \in \Gamma \cap A^2$, then $b \vee a = c \vee a$ for some $a \in J$. Therefore, $\langle a, 0\rangle \in \Delta$, so that $\langle b \vee a, b\rangle = \langle b \vee a, b \vee 0\rangle \in \Delta$ and $\langle b \vee a, c\rangle = \langle c \vee a, c\rangle = \langle c \vee a, c \vee 0\rangle \in \Delta$. Thus, $\langle b, c\rangle \in \Delta$. This shows that $\Gamma \cap A^2 \subseteq \Delta$. On the other hand, if $\langle b, c\rangle \in \Delta$, then $\langle b \wedge c', 0\rangle = \langle b \wedge c', c \wedge c'\rangle \in \Delta$ and $\langle b' \wedge c, 0\rangle = \langle b' \wedge c, b' \wedge b\rangle \in \Delta$. Hence $(b \wedge c') \vee (b' \wedge c) \in J$. Moreover, $b \vee ((b \wedge c') \vee (b' \wedge c)) = b \vee (b' \wedge c) = b \vee c$ and similarly $c \vee ((b \wedge c') \vee (b' \wedge c)) = c \vee b$. Thus, $\langle b, c\rangle \in \Gamma \cap A^2$. This completes the proof that $\Gamma \cap A^2 = \Delta$.

We will now show that the two element Boolean algebra **2** is injective in $\mathfrak{B}$. Let **A** and **B** be Boolean algebras, $\vartheta \in \text{Hom}(\mathbf{A}, \mathbf{B})$, $\varphi \in \text{Hom}(\mathbf{A}, \mathbf{2})$, where ϑ is a monomorphism. By Proposition 1.4.10 and 1.2.4, $\vartheta(A)$ determines a subalgebra of **B** that is isomorphic to **A**. Moreover, $\Delta = \{\langle \vartheta(a), \vartheta(b)\rangle \mid \varphi(a) = \varphi(b)\}$ is a congruence relation on this subalgebra. By Lemma 3.7, there is a congruence relation $\Gamma \in \Theta(\mathbf{B})$ such that $\Gamma \cap \vartheta(A)^2 = \Delta$. Note that $\Gamma \neq U_{\vartheta(A)}$, since $\Delta \neq \vartheta(A)^2$. (In fact $\varphi(0)$ $0 \neq 1 = \varphi(u)$, so that $\langle \vartheta(0), \vartheta(u)\rangle \notin \Delta$.) By the result of Example 2.4.2, there is a homomorphism χ of $\mathbf{B}/\Gamma$ to **2**. Let $\psi = \chi \circ \bar{\Gamma}$. If $a \in A$, and $\varphi(a) = 0$, then $\langle \vartheta(a), 0\rangle \in \Delta \subseteq \Gamma$, so that $\psi(\vartheta(a)) = \chi(\bar{\Gamma}(\vartheta(a))) = 0$. Similarly, if $\varphi(a) = u$, then $\psi(\vartheta(a)) = u$. Hence, $\psi \circ \vartheta = \varphi$. This proves the claim that **2** is injective in $\mathfrak{B}$. By Lemma 3.6, every direct product of copies of **2** is injective. By Example 2.4.2, every Boolean algebra is isomorphic to a subalgebra of such a product, so that $\mathfrak{B}$ is injectively complete.

Problems

1. Let $\mathfrak{A}$ be a class of partial algebras of type τ. Let $\mathbf{A}$ be a partial algebra of type τ. Suppose that $\mathbf{B}$ is a free $\mathfrak{A}$-extension of $\mathbf{A}$.

(a) Prove that $\mathbf{B}$ is a free $\mathscr{I}\mathfrak{A}$-extension of $\mathbf{A}$.

(b) Prove that $\mathbf{B}$ is a free $\mathscr{S}\mathfrak{A}$-extension of $\mathbf{A}$.

(c) Prove that $\mathbf{B}$ is a free $\mathscr{P}\mathfrak{A}$-extension of $\mathbf{A}$.

2. (a) Let $\mathfrak{A}$ be a class of partial algebras of type τ. Let $\mathbf{A}$ be a free partial $\mathfrak{A}$-algebra on the generating set X. Show that $\mathbf{A}$ is a free partial $\mathscr{Q}\mathfrak{A}$-algebra on the set X.

(b) Assume that $\mathscr{S}\mathfrak{A} = \mathscr{P}\mathfrak{A} = \mathfrak{A}$, and $\mathfrak{A}$ contains partial algebras with more than one element. Let X be any set. Prove that a partial algebra $\mathbf{A}$ is a free partial $\mathfrak{A}$-algebra on the generating set X if and only if $\mathbf{A}$ is a free partial $\mathscr{Q}\mathfrak{A}$-algebra on the set X.

3. The purpose of this problem is to outline a proof of the following generalization of Theorem 2.5.

Theorem. Let $\mathfrak{A}$ be a class of partial algebras of type τ. Suppose that $\mathscr{S}\mathfrak{A} = \mathscr{P}\mathfrak{A} = \mathfrak{A}$. Let $\{\mathbf{A}_i \mid i \in I\}$ be a nonempty set of partial algebras of type τ, which has the embedding property (EP): there exists $\mathbf{B} \in \mathfrak{A}$ and monomorphisms $\psi_i \in \operatorname{Hom}(\mathbf{A}_i, \mathbf{B})$. Then the free $\mathfrak{A}$-sum of $\{\mathbf{A}_i \mid i \in I\}$ exists.

(a) Let $\{\mathbf{B}_i \mid i \in I\}$ be a set of partial algebras of type τ, where $\mathbf{B}_i = \langle B_i; G_{\xi i}\rangle_{\xi<\rho}$. Assume that $B_i \cap B_j = \varnothing$ if $i \neq j$. Suppose also that if $\tau(\xi) = 0$, then $G_{\xi i} = \varnothing$ for each $i \in I$. Let $B = \bigcup_{i\in I} B_i$, $G_\xi = \bigcup_{i\in I} G_{\xi i}$, and $\mathbf{B} = \langle B; G_\xi\rangle_{\xi<\rho}$. Prove that $\mathbf{B}$ is a partial algebra of type τ, and $\langle \mathbf{B}; I_{B_i}\rangle_{i\in I}$ is a free sum of $\{\mathbf{B}_i \mid i \in I\}$ in the class of all partial algebras of type τ. Use this result to prove that if the partial algebras $\mathbf{A}_i = \langle A_i; F_{\xi i}\rangle_{\xi<\rho}$ (given in the theorem) satisfy the condition: $\tau(\xi) = 0$ implies $F_{\xi i} = \varnothing$, then the free sum of $\{\mathbf{A}_i \mid i \in I\}$ exists in the class of all partial algebras of type τ.

(b) For $i \in I$, define $\mathbf{A}_1^* = \langle A_i; F_{\xi 1}^*\rangle_{\xi<\rho}$, where $F_{\xi i}^* = F_{\xi i}$ if $\tau(\xi) > 0$ and $F_{\xi i}^* = \varnothing$ if $\tau(\xi) = 0$. Let $\langle \mathbf{A}^*; \lambda_i^*\rangle_{i\in I}$ be a free sum of $\{\mathbf{A}_i^* \mid i \in I\}$ in the class of all partial algebras of type τ. Let Γ be the intersection of all congruences in $\Theta(\mathbf{A}^*)$ that are of the form Γ_φ, where $\varphi \in \operatorname{Hom}(\mathbf{A}^*, \mathbf{C})$ is such that $\varphi \circ \lambda_i^* \in \operatorname{Hom}(\mathbf{A}_i, \mathbf{C})$ for all $i \in I$. Prove that $\langle A^*/\Gamma; \bar{\Gamma} \circ \lambda_i^*\rangle_{i\in I}$ is a free product of $\{\mathbf{A}_i \mid i \in I\}$ in the class of all partial algebras of type τ. (The proof will use the assumption that $\{\mathbf{A}_i \mid i \in I\}$ satisfies (EP).)

(c) Arguing as in last part of the proof of Theorem 2.5, show that the result of **(b)** leads to the theorem stated at the beginning of this problem.

4. Let $\mathfrak{A}$ be a class of partial algebras of type τ such that $\mathscr{S}\mathfrak{A} = \mathscr{P}\mathfrak{A} = \mathfrak{A}$. If $\{\mathbf{A}_0, \mathbf{A}_1, \cdots, \mathbf{A}_{k-1}\}$ are partial algebras of type τ, let $\mathbf{A}_0 \oplus \mathbf{A}_1 \oplus \cdots \oplus \mathbf{A}_{k-1}$ denote a free $\mathfrak{A}$-sum of $\{\mathbf{A}_0, \mathbf{A}_1, \cdots, \mathbf{A}_{k-1}\}$ whenever this set satisfies (EP). That is if $\langle \mathbf{B}; \sigma_0, \sigma_1, \cdots, \sigma_{k-1}\rangle$ is a free $\mathfrak{A}$-sum of $\{\mathbf{A}_0, \mathbf{A}_1, \cdots, \mathbf{A}_{k-1}\}$, let $\mathbf{B}$ be denoted by $\mathbf{A}_0 \oplus \mathbf{A}_1 \oplus \cdots \oplus \mathbf{A}_{k-1}$. Of course, $\mathbf{A}_0 \oplus \mathbf{A}_1 \oplus \cdots \oplus \mathbf{A}_{k-1}$ is determined only up to isomorphism.

(a) Show that if $\mathbf{A}_0 \oplus \mathbf{A}_1 \oplus \cdots \oplus \mathbf{A}_{k-1}$ exists, then $\mathbf{A}_{i_0} \oplus \cdots \oplus \mathbf{A}_{i_{r-1}}$ exists for $\{i_0, \cdots, i_{r-1}\} \subseteq k$, where $\{i_0, \cdots, i_r\}$ are not necessarily distinct. In particular, prove that if $\mathbf{A} \in \mathfrak{A}$, then $n\mathbf{A} = \mathbf{A} \oplus \mathbf{A} \oplus \cdots \oplus \mathbf{A}$ (n factors) exists.

(b) Prove that $(\mathbf{A} \oplus \mathbf{B}) \oplus \mathbf{C} \cong \mathbf{A} \oplus (\mathbf{B} \oplus \mathbf{C}) \cong \mathbf{A} \oplus \mathbf{B} \oplus \mathbf{C}$.

(c) Prove that if $\mathbf{A} \cong \mathbf{B}$ and $\mathbf{C} \cong \mathbf{D}$, then $\mathbf{A} \oplus \mathbf{C} \cong \mathbf{B} \oplus \mathbf{D}$.

(d) Let $\mathbf{A}$ be in $\mathfrak{A}$. Prove that either the partial algebras $n\mathbf{A}$, $n = 1, 2, 3, \cdots$, are pairwise nonisomorphic, or else there exist positive integers r and s such that $m\mathbf{A} \cong n\mathbf{A}$ for $m \neq n$ if and only if $m \geqq r$, $n \geqq r$, and $m \equiv n (\text{mod } s)$.

5. Let $\mathfrak{A}$ be a class of partial algebras of type $\tau \in (\text{Ord})^\rho$ satisfying $\mathscr{S}\mathfrak{A} = \mathscr{P}\mathfrak{A} = \mathfrak{A}$, such that $\mathfrak{A}$ contains algebras with at least two elements.

(a) Let $\mathbf{A}$ and $\mathbf{B}$ be free partial $\mathfrak{A}$-algebras on free generating sets, X and Y, respectively. Prove that if $|X| = |Y|$, then $\mathbf{A} \cong \mathbf{B}$.

(b) Let $\mathbf{A} \in \mathfrak{A}$, $\mathbf{B} \in \mathfrak{A}$, and let $X \subseteq A$ be such that $[X] = A$. Prove that $|\text{Hom}(\mathbf{A}, \mathbf{B})| \leqq |B^X|$, and that equality holds if $\mathbf{A}$ is free and X is a set of free generators of $\mathbf{A}$.

(c) Suppose that $\tau \in \omega^\rho$, where $\rho \leqq \omega$. Let $\mathbf{A}$ be a free partial $\mathfrak{A}$-algebra, and suppose that X and Y are sets of free generators of $\mathbf{A}$. Prove that if either X or Y is infinite, then $|X| = |Y|$. (*Hint:* use 1.3 if A is uncountable, and **(b)** if A is countable.)

(d) Let $\mathbf{A}$ be a free partial $\mathfrak{A}$-algebra, and suppose that X and Y are finite sets of free generators of $\mathbf{A}$. Prove that if there exists $\mathbf{B} \in \mathfrak{A}$ with $|B| < \aleph_0$, then $|X| = |Y|$.

(e) Let $\mathbf{A}$ be a free $\mathfrak{A}$-algebra. Use Example 2.3 and the result **(d)** of Problem **4** to show that the set of all natural numbers n for which there exists a set X of free generators of $\mathbf{A}$ such that $|X| = n$ is either empty, contains a single number, or else is an arithmetic progression.

6. Let $\mathfrak{A}$ be the class of all algebras of type $\{\langle 0, 2\rangle, \langle 1, 1\rangle, \langle 2, 1\rangle\}$ satisfying the identities $F_1(F_0(x, y)) = x$, $F_2(F_0(x, y)) = y$, $F_0(F_1(x), F_2(x)) = x$. Note that $\mathscr{S}\mathfrak{A} = \mathscr{P}\mathfrak{A} = \mathfrak{A}$.

(a) Define operations $F_0 \subseteq \omega^3$ by $F_0(m, n) = \frac{1}{2}((m + n)(m + n + 1)) + m$ and $F_1 \subseteq \omega^2$ and $F_2 \subseteq \omega^2$ by

$$F_1(n) = n - \tfrac{1}{2}(G(n)^2 + G(n)) \quad \text{and} \quad F_2(n) = \tfrac{1}{2}(G(n)^2 + 3G(n)) - n,$$

where $G(n)$ is the greatest integer $\leqq \frac{1}{2}((\sqrt{8n + 1}) + 1)$. Prove that $\langle \omega; F_0, F_1, F_2\rangle \in \mathfrak{A}$.

(b) Let $\mathbf{B} = \langle B; G_0, G_1, G_2 \rangle$ be the free $\mathfrak{A}$-algebra with the free generating set $\{x_0, x_1\}$ (with $x_0 \neq x_1$). Prove that $\{G_0(x_0, x_1)\}$ is a set of free generators for $\mathbf{B}$.

(c) Deduce from **(b)** that for every natural number n, the free $\mathfrak{A}$-algebra with n generators is isomorphic to the free $\mathfrak{A}$-algebra with one generator.

7. Let $\mathbf{S} = \langle S; \leqq \rangle$ be a partially ordered set. Define mappings $X \to X^*$ and $X \to X^*$ of $\mathbf{P}(S)$ to $\mathbf{P}(S)$ by

$$X^* = \{s \in S \mid x \leqq s \text{ for all } x \in X\} \qquad \text{and}$$
$$X_* = \{s \in S \mid s \leqq x \text{ for all } x \in X\}.$$

(a) Prove that $X \subseteq Y$ implies $X^* \supseteq Y^*$ and $X_* \supseteq Y_*$.
(b) Prove that $(X^*)_* \supseteq X$.
(c) Prove that $((X^*)_*)^* = X^*$.
(d) Prove that if $\{X_i \mid i \in I\} \subseteq \mathbf{P}(S)$, then

$$((\textstyle\bigcap_{i\in I}(X_i^*)_*)^*)_* = \textstyle\bigcap_{i\in I}(X_i^*)_*.$$

(e) Prove that if $s \in S$, then $(\{s\}^*)_* = \{x \in S \mid x \leqq s\}$. A subset N of S is called *normal* if $(N^*)_* = N$.

(f) Prove that the collection of all normal subsets of S, partially ordered by inclusion, is a complete lattice $\mathbf{N}(\mathbf{S})$, in which the g.l.b. and l.u.b. of $\{N_i \mid i \in I\}$ are $\bigcap_{i\in I} N_i$ and $((\bigcup_{i\in I} N_i)^*)_*$. $\mathbf{N}(\mathbf{S})$ is called the *normal completion* of $\mathbf{S}$.

(g) Show that the mapping $s \to N(s) = \{x \in S \mid x \leqq s\}$ preserves least upper and greatest lower bounds whenever they exist in $\mathbf{S}$. That is, if $\{s_i \mid i \in I\} \subseteq S$, and if $t \in S$ is the least upper (greatest lower) bound of $\{s_i \mid i \in I\}$ in $\mathbf{S}$, then $N(t)$ is the least upper (and respectively, greatest lower) bound of $\{N(s_i) \mid i \in I\}$ in $\mathbf{N}(\mathbf{S})$.

(h) Use Theorem 1.6 and the result **(g)** to prove that every partially ordered set $\mathbf{S}$ can be extended to a lattice $\mathbf{L}$ such that

(i) If $t = r \vee s$ in $\mathbf{S}$, then $t = r \vee s$ in $\mathbf{L}$.
(ii) If $t = r \wedge s$ in $\mathbf{S}$, then $t = r \wedge s$ in $\mathbf{L}$.
(iii) If φ is a mapping of $\mathbf{S}$ to a lattice $\mathbf{M}$ such that $\varphi(r \vee s) = \varphi(r) \vee \varphi(s)$ when $r \vee s$ exists in $\mathbf{S}$ and $\varphi(r \wedge s) = \varphi(r) \wedge \varphi(s)$ when $r \wedge s$ exists in $\mathbf{S}$, then there is a lattice homomorphism ψ of $\mathbf{L}$ into $\mathbf{M}$ that extends φ.

8. Let $\mathfrak{A}$ be a class of similar algebras such that for every cardinal number α, there exists a free $\mathfrak{A}$-algebra on a generating set of cardinality $\geqq \alpha$. Prove that $\mathscr{S}\mathfrak{A} \subseteq \mathscr{Q}\mathfrak{A}$.

9. Let $\{\mathbf{A}_i \mid i \in I\}$ be a set of partial algebras of type $\tau \in \alpha^o$. Assume that $\mathscr{F} \subseteq \mathbf{P}(I)$ satisfies

(i) $U \supseteq V$, $V \in \mathscr{F}$ implies $U \in \mathscr{F}$.
(ii) If $\zeta < \alpha$ and $\{U_\xi \mid \xi < \zeta\} \subseteq \mathscr{F}$, then $\bigcap_{\xi<\zeta} U_\xi \in \mathscr{F}$. Let $\mathbf{A} = \Pi_{i\in I}\mathbf{A}_i$, and define $\Gamma_{\mathscr{F}} \subseteq A^2$ by $\langle a, b \rangle \in \Gamma_{\mathscr{F}}$ if and only if $\{i \in I \mid a(i) = b(i)\} \in \mathscr{F}$.

(a) Prove that $\Gamma_{\mathscr{F}}$ is a congruence relation on $\mathbf{A}$. The *reduced direct product* of $\{\mathbf{A}_i \mid i \in I\}$ relative to $\mathscr{F}$ is defined to be $\mathbf{A}/\Gamma_{\mathscr{F}}$. It is convenient to denote this reduced product by $\Pi_{i \in I}\mathbf{A}_i/\mathscr{F}$.

(b) For $U \in \mathscr{F}$, denote $\mathscr{F} \upharpoonright U = \{V \cap U \mid V \in \mathscr{F}\}$. Show that $\mathscr{F} \upharpoonright U$ satisfies the conditions (i) and (ii) with I replaced by U. Prove that the mapping $\pi_{U,I} : \Pi_{i \in I}\mathbf{A}_i \to \Pi_{i \in U}\mathbf{A}_i$ (defined in Problem **10** of Chapter 1) induces a unique isomorphism $\lambda_{U,I} : \Pi_{i \in I}\mathbf{A}_i/\mathscr{F} \to \Pi_{i \in U}A_i/(\mathscr{F} \upharpoonright U)$ such that

$$\begin{array}{ccc} \Pi_{i \in I}\mathbf{A}_i & \xrightarrow{\pi_{U,I}} & \Pi_{i \in U}\mathbf{A}_i \\ \downarrow & & \downarrow \\ \Pi_{i \in I}\mathbf{A}_i/\mathscr{F} & \xrightarrow[\lambda_{U,I}]{} & \Pi_{i \in U}\mathbf{A}_i/(\mathscr{F} \upharpoonright U) \end{array}$$

is consistent (where the vertical maps are the epimorphisms $\Gamma_{\mathscr{F}}$ and $\Gamma_{\mathscr{F} \upharpoonright U}$).

Reduced products have many applications in algebra. The next problem shows one way in which they can be used.

10. Let α be a regular cardinal number. Suppose that $\tau \in \alpha^\rho$. Let I be a partially ordered set such that if $J \subseteq I$ and $|J| < \alpha$, then there is an $i \in I$ that satisfies $j \leqq i$ for all $j \in J$. Let $\mathscr{F}$ be the collection of all $U \in \mathbf{P}(I)$ that contain a set of the form $\{j \in I \mid j \geqq i\}$ for some $i \in I$.

(a) Prove that $\mathscr{F}$ satisfies the conditions (i) and (ii) of Problem **9**.

Let $\{\mathbf{A}_i \mid i \in I\}$ be a set of partial algebras of type τ. Suppose that a set $\{\varphi_{ij} \mid i, j \in I, j \leqq i\}$ of homomorphisms is given such that $\varphi_{ij} \in \text{Hom}(\mathbf{A}_j, \mathbf{A}_i)$ $\varphi_{ii} = I_{A_i}$, and $\varphi_{ij} \circ \varphi_{jk} = \varphi_{ik}$ for $i \leqq j \leqq k$.

(b) Prove that there is a set $\{\varphi_i \mid i \in I\}$ of homomorphisms, $\varphi_i \in \text{Hom}(\mathbf{A}_i, \Pi_{j \in I}\mathbf{A}_j/\mathscr{F})$, satisfying $\varphi_i \circ \varphi_{ij} = \varphi_j$ for $j \leqq i$. Moreover, if all φ_{ij} are monomorphisms, then so are all φ_i.

(c) Use the result of **(b)** to prove that if $\mathfrak{A}$ is a class of algebras of type τ, then $\mathscr{S}_\alpha^{-1}\mathfrak{A} \subseteq \mathscr{Q}\mathscr{S}\mathscr{P}\mathfrak{A}$ (see Problem **14** in Chapter 1).

11. Let $\mathfrak{A}$ be a class of partial algebras of type $\tau \in \omega^\rho$, such that $\mathscr{Q}\mathfrak{A} = \mathscr{S}\mathfrak{A} = \mathscr{P}\mathfrak{A} = \mathfrak{A}$. Assume that $\mathfrak{A}$ has the following *weak embedding property* (WEP). If $\mathbf{A} \in \mathfrak{A}$ and $\mathbf{B} \in \mathfrak{A}$, then there exists $\mathbf{C} \in \mathfrak{A}$ and monomorphisms $\varphi \in \text{Hom}(\mathbf{A}, \mathbf{C})$ and $\psi \in \text{Hom}(\mathbf{B}, \mathbf{C})$.

(a) Prove that every subset of $\mathfrak{A}$ has the embedding property (see Theorem 2.5). (*Hint:* use (WEP) and the result of Problem **10** to construct a partial algebra into which every partial algebra of a given set can be mapped monomorphically.)

(b) Let $\tau \in \omega^\rho$, where $\rho \leqq \omega$. Prove that if α is any infinite cardinal number, there exists $\mathbf{A} \in \mathfrak{A}$ such that

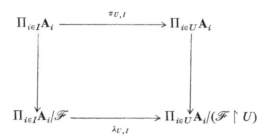

(i) $|A| \leqq 2^\alpha$.

(ii) If $\mathbf{B} \in \mathfrak{A}$ and $|B| \leqq \alpha$, then there is a monomorphism φ of $\mathbf{B}$ to $\mathbf{A}$.

12. Let $\mathfrak{A}$ be a class of partial algebras such that $\mathscr{Q}\mathfrak{A} = \mathscr{S}\mathfrak{A} = \mathscr{P}\mathfrak{A} = \mathfrak{A}$. Assume that the class $\mathfrak{A}$ satisfies the following *amalgamation property* (AP). If $\mathbf{A}$, $\mathbf{B}$, and $\mathbf{C}$ belong to $\mathfrak{A}$, and there exist monomorphisms $\varphi \in \operatorname{Hom}(\mathbf{A}, \mathbf{B})$ and $\psi \in \operatorname{Hom}(\mathbf{A}, \mathbf{C})$, then there exists $\mathbf{D} \in \mathfrak{A}$ and $\varphi' \in \operatorname{Hom}(\mathbf{B}, \mathbf{D})$ and $\psi' \in \operatorname{Hom}(\mathbf{C}, \mathbf{D})$ such that $\varphi' \circ \varphi = \psi' \circ \psi$.

(a) Let $\{\mathbf{A}_\eta \mid \eta < \alpha\} \subseteq \mathfrak{A}$ where α is a limit ordinal, and suppose that $\{\mathbf{B}_\eta \mid \eta < \alpha\}$ is a set of subalgebras of a partial algebra $\mathbf{B} \in \mathfrak{A}$ such that monomorphisms $\mu_\eta : \mathbf{B}_\eta \to \mathbf{A}_\eta$ exist for all $\eta < \alpha$. Prove that there is a partial algebra $\mathbf{C} \in \mathfrak{A}$ that is an extension of $\mathbf{B}$ and a set ν_η of monomorphisms of $\mathbf{A}_\eta$ to $\mathbf{C}$ such that $\nu_\eta \circ \mu_\eta = I_{B_\eta}$ for all $\eta < \alpha$. (*Hint:* use (AP) and the result of Problem **10** to construct a well-ordered sequence of extensions $\mathbf{C}_0 = \mathbf{B}, \mathbf{C}_1, \cdots, \mathbf{C}_\eta, \cdots, \eta < \alpha$ such that there is a monomorphism ν_η of $\mathbf{A}_\eta$ to $\mathbf{C}_{\eta+1}$ satisfying $\nu_\eta \circ \mu_\eta = I_{B_\eta}$. Then $\mathbf{C} = \mathbf{C}_\alpha$ is the required partial algebra.)

(b) Use (a) to prove that if $\mathfrak{A}$ satisfies (AP), then free sums with an amalgamated subalgebra exist in $\mathfrak{A}$.

(c) Assume that $\mathfrak{A}$ is a class of abstract algebras that satisfies (AP) and $\mathscr{Q}\mathfrak{A} = \mathscr{S}\mathfrak{A} = \mathscr{P}\mathfrak{A} = \mathfrak{A}$. Let $\{\mathbf{A}_i \mid i \in I\} \subseteq \mathfrak{A}$, and suppose that for each $i \in I$, $\mathbf{B}_i$ is a subalgebra of $\mathbf{A}_i$. Let $\langle \mathbf{A}; \sigma_i \rangle_{i \in I}$ be a free $\mathfrak{A}$-sum of $\{\mathbf{A}_i \mid i \in I\}$. Define $\mathbf{B} = \mathbf{A} \upharpoonright B$, where $B = [\bigcup_{i \in I} \sigma_i(B_i)]_{\mathbf{A}}$. Prove that $\langle \mathbf{B}; \sigma_i \upharpoonright B_i \rangle_{i \in I}$ is a free $\mathfrak{A}$-sum of $\{\mathbf{B}_i \mid i \in I\}$. (*Hint:* note that the existence of a free $\mathfrak{A}$-sum of $\{\mathbf{A}_i \mid i \in I\}$ implies that $\mathbf{K}(\mathbf{A}_i) \cong \mathbf{K}(\mathbf{A}_j)$ for all i and j in I. From this observation and **(b)** deduce that $\{\mathbf{B}_i \mid i \in I\}$ has a free $\mathfrak{A}$-sum $\langle \mathbf{B}'; \tau_i \rangle_{i \in I}$, and there is a homomorphism σ of $\mathbf{B}'$ onto $\mathbf{B}$ satisfying $\sigma \circ \tau_i = \sigma_i \upharpoonright B_i$. Use **(a)** to obtain an algebra $\mathbf{C} \in \mathfrak{A}$, which is an extension of $\mathbf{B}'$, and monomorphisms ν_i of $\mathbf{A}_i$ to $\mathbf{C}$. Deduce that there is a homomorphism ν of $\mathbf{A}$ to $\mathbf{C}$ such that $\nu \circ \sigma = I_{B'}$. Thus, σ is an isomorphism and the result follows.)

(d) With the same hypotheses as in **(c)**, let $\mathfrak{U}$ be the class of all subalgebras of free $\mathfrak{A}$-algebras. Prove that $\mathfrak{U}$ satisfies $\mathscr{S}\mathfrak{U} = \mathfrak{U}$, and any free $\mathfrak{A}$-sum of algebras in $\mathfrak{U}$ is in $\mathfrak{U}$.

13. Let $\mathbf{R} = \langle R; P \rangle$ and $\mathbf{S} = \langle S; Q \rangle$ be partially ordered sets (that is, $P \subseteq R^2$ and $Q \subseteq S^2$ satisfy the conditions for a partial ordering). Suppose that $P \cap (R \cap S)^2 = Q \cap (R \cap S)^2$. Let $T = R \cup S$ and $M = P \cup Q \cup (P \circ Q) \cup (Q \circ P)$ (see 3.1.3).

(a) Prove that $\mathbf{T} = \langle T; M \rangle$ is a partially ordered set, and that $M \cap R^2 = P$, $M \cap (R \times S) = P \circ Q$, $M \cap (S \times R) = Q \circ P$, and $M \cap S^2 = Q$.

(b) Show that if $a = b \vee c$ $(a = b \wedge c)$ in $\mathbf{R}$, then $a = b \vee c$ (and respectively, $a = b \wedge c$) in $\mathbf{T}$.

(c) Use these results, together with the conclusion of Problem **7**, to show that the class of all lattices satisfies the condition (AP) of Problem **12**, so that free sums with amalgamation exist in the class of all lattices.

14. Let $\mathfrak{A}$ be a class of partial algebras of type τ. A partial algebra **F** of type τ is called *projective* for $\mathfrak{A}$ if whenever **A** and **B** are in $\mathfrak{A}$, $\varphi \in \text{Hom}(\mathbf{F}, \mathbf{B})$ and $\vartheta \in \text{Hom}(\mathbf{A}, \mathbf{B})$ with ϑ an epimorphism, there exists $\psi \in \text{Hom}(\mathbf{F}, \mathbf{A})$ such that $\vartheta \circ \psi = \varphi$. That is,

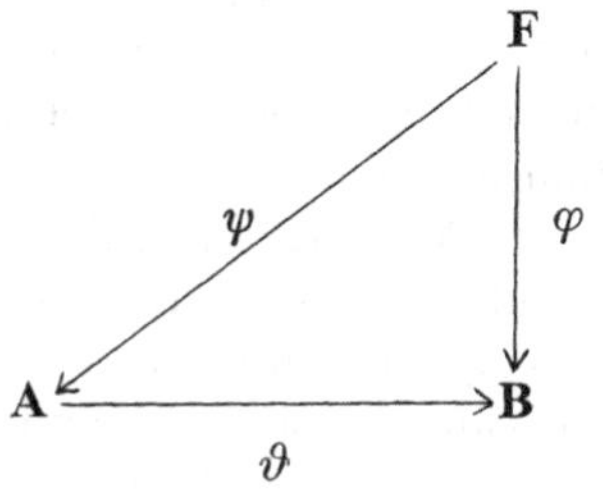

is consistent. If $\mathbf{F} \in \mathfrak{A}$, then **F** is called *projective in* $\mathfrak{A}$.

(a) Let **F** be projective for $\mathfrak{A}$, and suppose that **G** is a free $\mathfrak{A}$-extension of **F**. Show that **G** is projective in $\mathfrak{A}$.

(b) Prove that every free $\mathfrak{A}$-algebra is projective in $\mathfrak{A}$.

(c) A class $\mathfrak{A}$ of partial algebras of type τ is called *projectively complete* if every $\mathbf{A} \in \mathfrak{A}$ is an epimorphic image of a partial algebra which is projective in $\mathfrak{A}$. Prove that if $\mathscr{S}\mathfrak{A} = \mathscr{P}\mathfrak{A} = \mathfrak{A}$, then $\mathfrak{A}$ is projectively complete.

(d) Suppose that $\mathfrak{A}$ is a class of abstract algebras that satisfies $\mathscr{S}\mathfrak{A} = \mathscr{P}\mathfrak{A} = \mathfrak{A}$. Prove that an algebra **F** is projective in $\mathfrak{A}$ if and only there is a free $\mathfrak{A}$-algebra **K** and homomorphisms $\lambda \in \text{Hom}(\mathbf{K}, \mathbf{F})$ and $\mu \in \text{Hom}(\mathbf{F}, \mathbf{K})$ such that $\lambda \circ \mu = I_F$.

15. Let $\mathfrak{A}$ be a class of partial algebras of type τ such that $\mathscr{P}\mathfrak{A} = \mathfrak{A}$. Let $\mathfrak{D}$ be the class of all partial algebras that are subalgebras of a partial algebra, which is injective in $\mathfrak{A}$. Prove that $\mathscr{S}\mathfrak{D} = \mathscr{P}\mathfrak{D} = \mathfrak{D}$, and $\mathfrak{D}$ is injectively complete.

16. Let $\mathfrak{A}$ be a class of partial algebras of type τ such that $\mathscr{S}\mathfrak{A} = \mathfrak{A}$.

(a) Let **A** be a subalgebra of a partial algebra **D** which is injective in $\mathfrak{A}$. Prove that **A** is injective in $\mathfrak{A}$ if and only if there exists $\varphi \in \text{Hom}(\mathbf{D}, \mathbf{A})$ such that $\varphi \restriction A = I_A$. In this case, **A** is called a *retract* of **D**.

(b) Show that a Boolean algebra **B** is injective if and only if **B** is isomorphic to a retract of the Boolean algebra of all subsets of some set X.

(c) Use **(b)** to prove that every injective Boolean algebra is complete.

(d) Let **B**, **C**, and **D** be Boolean algebras with **B** a subalgebra of **C** and **C** a subalgebra of **D**. Assume that $a \in D$ and $D = [C \cup \{a\}]$ in $\Sigma(\mathbf{D})$. Show that $D = \{(a \wedge c) \vee (a' \wedge d) \mid c, d \in C\}$. Assume that $\varphi \in \text{Hom}(\mathbf{C}, \mathbf{B})$. Let $b = \bigvee\{\varphi(x) \mid x \in C, x \leq a\}$ in B. Show that $\psi((a \wedge c) \vee (a' \wedge d)) = (b \wedge \varphi(c)) \vee (b' \wedge \varphi(d))$ for c, d in C gives a well-defined mapping ψ of D to B such that ψ is an extension of φ to a homomorphism of **D** to **B**. (*Hint:* the proof that ψ is well defined requires the infinite distributive law $u \wedge (\bigvee v_i) = \bigvee(u \wedge v_i)$. Show that this law holds in every complete Boolean algebra.)

(e) Use Zorn's lemma and parts **(b)** and **(d)** above to prove that every complete Boolean algebra is injective in the class of all Boolean algebras.

17. Let $\mathfrak{A}$ be a class of partial algebras of type τ such that $\mathscr{Q}\mathfrak{A} = \mathscr{S}\mathfrak{A} = \mathfrak{A}$ and $\mathfrak{A}$ is injectively complete.

(a) Show that if $\mathbf{A}$ is a subalgebra of $\mathbf{B}$ where $\mathbf{B}$ is in $\mathfrak{A}$, then $\Gamma \to \Gamma \cap A^2$ maps $\Theta(\mathbf{B})$ onto $\Theta(\mathbf{A})$.

(b) Prove that every subalgebra of a simple|| partial algebra in $\mathfrak{A}$ is simple.

(c) Use **(b)** to show that the following classes are not injectively complete: the class of all groups, the class of all rings, and the class of all lattices.

18. Following the outline of Example 3.5, show that the class of all distributive lattices is injectively complete. (*Hint:* because of the result of Problem **13** in Chapter 2, the crucial step is the proof that if $\mathbf{L}$ is a sublattice of $\mathbf{M}$, then any homomorphism of $\mathbf{L}$ to $\mathbf{2}$ can be extended to a homomorphism of $\mathbf{M}$ to $\mathbf{2}$. As in 3.5, it suffices to show that if $\Delta \in \Theta(\mathbf{L})$ and $\mathbf{L}/\Delta \cong \mathbf{2}$, then there exists $\Gamma \in \Theta(\mathbf{M})$ such that $\Gamma \cap L^2 = \Delta$. The same construction that was given in 3.5 will work in this case, even though $\mathbf{L}$ may not have a zero element.)

19. Show that the class of all semilattices is injectively complete.

Notes to Chapter 4

The extension theorem 1.6 can be found in a slightly less general form in the work [23] of Grätzer and Schmidt. The proof which they give is constructive. Our proof is an adaptation of the argument which is used in Christensen and Pierce [8] to obtain the existence of free sums (see also [46]). The applications of 1.6 given in 1.7 and Problem **7** are useful in the study of free lattices. Dean [12] proved the existence of the extension given in 1.7, while the extension in Problem **7** was constructed by Dilworth [13]. A result substantially equivalent to Theorem 2.5 first occurs in Sikorski's paper [46]. Another existence theorem for free sums appears in [42]. The observation that free sums with amalgamation can be treated as a special case of ordinary free sums (by introducing constants) was made by Gaifman (unpublished).

Most of the principal results of this chapter can be obtained in the more general context of category theory. Such a program has been carried out by W. Felscher. This approach clearly exposes the duality between free sums and free products.

|| Recall that $\mathbf{B}$ is simple if $\Theta(\mathbf{B}) = \{I_B, U_B\}$.

The problem of finding conditions under which the number of generators of a free algebra is unique has received considerable attention. The elementary results that are indicated in Problems **5** and **6** come (mostly) from a paper of Jónsson and Tarski [30]. Swierczkowski [50] has shown that any arithmetic progression can occur as the set of cardinalities of generators of a suitable free algebra (see Problem **5(e)**). This work is related to extensive studies of independence in abstract algebra which was initiated by Marczewski (see [37] and [38]) and cultivated by Swierczkowski [49] and Urbanik ([54] and [55]).

An elementary exposition of the reduced products that are defined in Problem **9** can be found in the papers of Frayne, Morel, and Scott [19] and Kochen [33]. Recently this concept has been used profitably in the study of axiomatic model classes. An accessible introduction to this topic is given in Cohn's Chapters 5 and 6 (see [9]). This book also provides references to the basic literature on the subject.

Problem **12** hints at some of the work that has been done on the existence of universal relational systems. The initial published results on this topic are in Jónsson [27]. The most recent work on the subject appears in the paper of Morley and Vaught [39], which also contains a good bibliography.

The theorem sketched in Problem **16** is a celebrated result of Sikorski [45] and Halmos [24]. It has an important application in Halmos's theory of polyadic algebras.

[5]

Varieties of Algebras

1. Word Algebras

Most familiar classes of algebras such as groups and rings are defined by identities. For example, the collection of all groups is the class of algebras of type $\{\langle 0, 2\rangle, \langle 1, 1\rangle\}$ satisfying the identities

$$\mathbf{x}_0 \cdot (\mathbf{x}_1 \cdot \mathbf{x}_2) = (\mathbf{x}_0 \cdot \mathbf{x}_1) \cdot \mathbf{x}_2,$$
$$\mathbf{x}_0 \cdot (\mathbf{x}_1 \cdot \mathbf{x}_1^{-1}) = \mathbf{x}_0,$$
$$(\mathbf{x}_0 \cdot \mathbf{x}_0^{-1}) \cdot \mathbf{x}_1 = \mathbf{x}_1.$$

In this chapter we will study identities that are defined for abstract algebras of arbitrary type. (Our results will not apply to partial algebras.) Since the types under consideration are not assumed to be finitary, it is necessary to examine identities that involve infinitely many symbols.

Identities are always of the form $\mathbf{w}_0 = \mathbf{w}_1$, where $\mathbf{w}_0$ and $\mathbf{w}_1$ are sequences of symbols, called *terms* or *words*. For example, the words which occur in the identities for groups given above are $\mathbf{x}_0 \cdot (\mathbf{x}_1 \cdot \mathbf{x}_2)$, $(\mathbf{x}_0 \cdot \mathbf{x}_1) \cdot \mathbf{x}_2$, $\mathbf{x}_0 \cdot (\mathbf{x}_1 \cdot \mathbf{x}_1^{-1})$, $\mathbf{x}_0$, $(\mathbf{x}_0 \cdot \mathbf{x}_0^{-1}) \cdot \mathbf{x}_1$, and $\mathbf{x}_1$. This preliminary section presents a careful definition of the notion of *words of type* τ, and a discussion of the set of all such words.

Words in any language consist of sequences of symbols. Consequently, we begin by discussing sequences in general.

Let S be a nonempty set. Suppose that W is any well-ordered set. An element $\mathbf{s} \in S^W$ is called a *sequence of elements in S indexed by W*. In most cases W will be an ordinal number. However, it would be inconvenient to add this restriction to the definition of sequences. In practice, the nature of the indexing set is not important, and we will define sequences (as opposed to indexed sequences, that is, sequences indexed by some set) in such a way that no particular indexing set is favored over another. If $\mathbf{s}$ and $\mathbf{t}$ are sequences of elements in S indexed by well-ordered sets V and W respectively, then $\mathbf{s}$ and $\mathbf{t}$ will be called equivalent if there is an order isomorphism φ of V to W such that $\mathbf{s} = \mathbf{t} \circ \varphi$. This definition obviously describes an equivalence relation on the class of indexed sequences. We define a sequence of elements in S to be an equivalence class of indexed sequences under this equivalence relation. Of course, it is usually not necessary to preserve the formal distinction between "sequences" and "indexed sequences." We will frequently refer to "the sequence **s**," meaning either "the indexed sequence **s**," or else "the sequence consisting of all indexed sequences which are equivalent to **s**." Usually, either interpretation will be acceptable; otherwise, the interpretation will be clear from the context.

If **s** is a sequence indexed by the well-ordered set W, the order type of W is called the *length* of **s**, and is denoted by $\lambda(\mathbf{s})$. Thus,

$$\lambda(\mathbf{s}) = \tilde{W}$$

is a uniquely determined ordinal number. Equivalent indexed sequences have the same length, so that it makes sense to speak of the length of a sequence. It is clear that every indexed sequence **s** is equivalent to a unique sequence indexed by $\lambda(\mathbf{s})$.

Let **s** be a sequence indexed by W. Any indexed sequence **t** that is equivalent to a sequence of the form $\mathbf{s} \restriction V$ for some $V \subseteq W$ is called a *subsequence of* **s**. In particular, $\mathbf{s} \restriction V$ is a subsequence of **s**. Plainly, the relation of being a subsequence is invariant under equivalence. That is, if $\mathbf{s}_0$ is equivalent to $\mathbf{s}_1$ and $\mathbf{t}_0$ is equivalent to $\mathbf{t}_1$, then $\mathbf{t}_0$ is a subsequence of $\mathbf{s}_0$ if and only if $\mathbf{t}_1$ is a subsequence of $\mathbf{s}_1$. Therefore, the definition of subsequences of an indexed sequence gives rise to a notion of subsequences of a sequence.

A useful fact follows from these definitions and the observation that if $V \subseteq W$, where W is well ordered, then $\tilde{V} \leqq \tilde{W}$ (see 3.4 of the Introduction):

1.1 LEMMA. If **s** is a subsequence of **t**, then $\lambda(\mathbf{s}) \leqq \lambda(\mathbf{t})$.

Let S be any set, and let ν be an ordinal number. Suppose that **s** is a sequence of elements in $\bigcup_{\mu<\nu} S^\mu$ indexed by a well-ordered set V. That is, for each $v \in V$, $\mathbf{s}(v)$ is a sequence of elements in S indexed by some $\mu < \nu$ (where μ depends on v). Let $W = \{\langle v, \zeta\rangle \in V \times \nu \mid \zeta < \lambda(\mathbf{s}(v))\}$, where W has the ordering induced by the lexicographic order on $V \times \nu$. Thus, W is well ordered (see section 3 of the Introduction). The *concatenation* of **s** is defined to be the sequence **t** of elements in S indexed by W which is defined by

$$\mathbf{t}(\langle v, \zeta\rangle) = \mathbf{s}(v)(\zeta),$$

for $\langle v, \zeta\rangle \in W$. This definition of the concatenation of a sequence of *indexed* sequences leads to a definition of the concatenation of a sequence of sequences. Indeed, if **s** is a sequence of sequences indexed by the well-ordered set V, let $\nu = \bigcup_{v\in V}\lambda(\mathbf{s}(v))$, and for each $v \in V$, let $\mathbf{s}'(v)$ be the sequence indexed by $\lambda(\mathbf{s}(v))$, which belongs to the equivalence class $\mathbf{s}(v)$. If $\mathbf{t}'$ is the concatenation of $\mathbf{s}'$ (as defined above), then the sequence **t** which is the equivalence class of $\mathbf{t}'$ will be called the *concatenation of* **s**.

In most cases, a sequence of sequences (or sequence of indexed sequences) will be designated as a set of sequences with subscript notation: $\{\mathbf{s}_v \mid v \in V\}$ or $\{\mathbf{s}_\eta \mid \eta < \mu\}$. The concatenation will be denoted by the expressions

$$\langle \cdots \mathbf{s}_v \cdots \rangle_{v\in V} \quad \text{or} \quad \langle \mathbf{s}_0 \cdots \mathbf{s}_\eta \cdots \rangle_{\eta<\mu},$$

with more terms indicated if necessary. When V or μ is finite or countable, the notation can be simplified without loss of clarity. For example, we will use such expressions as

$$\mathbf{s}_0\mathbf{s}_1\mathbf{s}_2, \quad \mathbf{s}_0\mathbf{s}_1 \cdots \mathbf{s}_{n-1}, \quad \text{or} \quad \mathbf{s}_0\mathbf{s}_1 \cdots \mathbf{s}_n \cdots,$$

with the obvious meanings.

If s denotes an element of S, it is convenient to designate the one element sequence $\{\langle 0, s\rangle\}$ by the corresponding bold-faced letter $\mathbf{s}$. This convention enables us to specify a (short) finite sequence by simply writing it in boldface letters.

Suppose that $\{\mathbf{s}_v \mid v \in V\}$ is a sequence of sequences (or indexed sequences) that is indexed by the well-ordered set V. If $U \subseteq V$, then the concatenation of the subsequence $\{\mathbf{s}_u \mid u \in U\}$ is plainly a subsequence of the concatenation of $\{\mathbf{s}_v \mid v \in V\}$. In particular, $\lambda(\langle \cdots \mathbf{s}_u \cdots \rangle_{u \in U}) \leqq \lambda(\langle \cdots \mathbf{s}_v \cdots \rangle_{v \in V})$.

We turn now to the discussion of words of type $\tau \in \mathrm{Ord}^\rho$. The symbols occurring in these words will be of three types:

(1) Φ_ξ for $\xi < \rho$,
(2) x_ζ for $\zeta < \alpha$, where α is a fixed cardinal number, and
(3) the parentheses symbols (and).

The symbol Φ_ξ is to be interpreted as representing a $\tau(\xi)$-ary operation, and is called an *operation symbol.* The symbol x_ζ is called an *individual variable.* The set of all these symbols will be called the *standard alphabet for algebras of type τ with α variables*, and will be denoted $S_{\tau\alpha}$.

1.2 Definition. The *words* of type τ in α variables are the elements of the smallest set $W_{\tau\alpha}$ of sequences of elements in $S_{\tau\alpha}$ satisfying

(i) $\mathbf{x}_\zeta \in W_{\tau\alpha}$ for all $\zeta < \alpha$, and
(ii) if $\xi < \rho$ and $w_\eta \in W_{\tau\alpha}$ for each $\eta < \tau(\xi)$, then

$$\langle \boldsymbol{\Phi}_\xi(\mathbf{w}_0\mathbf{w}_1 \cdots \mathbf{w}_\eta \cdots)\rangle_{\eta < \tau(\xi)} \in W_{\tau\alpha}.$$

In order to justify this definition, it is necessary to establish the existence of a smallest set $W_{\tau\alpha}$ satisfying (i) and (ii) of 1.2. The desired result follows from two observations:

(a) If $\{W_i \mid i \in I\}$ is a set of sets satisfying (i) and (ii) of Definition 1.2, then $\bigcap_{i \in I} W_i$ satisfies these conditions.

(b) There exists a *set* W satisfying (i) and (ii) of Definition 1.2.

In fact, if W is a set satisfying 1.2 (i) and (ii), then it is evident that the intersection of all subsets of W, which satisfy these conditions, is contained in every set that fulfills the conditions of Definition 1.2.

The statement (a) is obvious. To prove (b), let β be the smallest infinite cardinal number such that $|\tau(\xi)| \leqq \beta$ for all $\xi < \rho$. It is then easy to see that the collection W of all sequences $\mathbf{s}$ of elements in $S_{\tau\alpha}$ such that $\lambda(\mathbf{s}) \leqq \beta$ has the properties (i) and (ii) of 1.2. Each $\mathbf{s} \in W$ is represented by a sequence indexed by an ordinal $\leqq \beta$, so that W is a set. This observation not only completes the justification of Definition 1.2, but it also shows that if $\mathbf{w} \in W_{\tau\alpha}$, then $\lambda(\mathbf{w}) \leqq \beta$. In case $\tau(\xi) < \omega$ for all $\xi < \rho$, a stronger conclusion is obtained. Indeed, in this case the set of all finite sequences of elements in $S_{\tau\alpha}$ has the properties listed in 1.2. Therefore, if τ is finitary, then $\lambda(\mathbf{w})$ is finite for all $\mathbf{w} \in W_{\tau\alpha}$.

1.3 LEMMA. Let $\mathbf{w} \in W_{\tau\alpha}$. Then $\lambda(\mathbf{w}) > 0$. If $\lambda(\mathbf{w}) = 1$, then $\mathbf{w} = \mathbf{x}_\zeta$ for some $\zeta < \alpha$. If $\lambda(\mathbf{w}) > 1$, then there is a $\xi < \rho$ and elements $\mathbf{w}_\eta \in W_{\tau\alpha}$ such that

$$\mathbf{w} = \langle \mathbf{\Phi}_\xi(\mathbf{w}_0\mathbf{w}_1 \cdots \mathbf{w}_\eta \cdots)\rangle_{\eta < \tau(\xi)}$$

and

$$\lambda(\mathbf{w}_\eta) < \lambda(\mathbf{w}) \qquad \text{for all} \qquad \eta < \tau(\xi).$$

Proof. Let W_0 be the class of all sequences $\mathbf{v}$ of elements in $S_{\tau\alpha}$ such that either $\mathbf{v} = \mathbf{x}_\zeta$ for some $\zeta < \alpha$, or else $\lambda(\mathbf{v}) > 1$. It is clear that W_0 satisfies the conditions (i) and (ii) of Definition 1.2, so that $W_{\tau\alpha} \subseteq W_0$. Since $\lambda(\mathbf{v}) > 0$ for all $\mathbf{v} \in W_0$, it follows that $\lambda(\mathbf{w}) > 0$. Moreover, if $\lambda(\mathbf{w}) = 1$, then $\mathbf{w} = \mathbf{x}_\zeta$ for some $\zeta < \alpha$. The last statement of the lemma is proved by a similar construction. Let W_1 be the class of all sequences $\mathbf{v}$ of elements in $S_{\tau\alpha}$ such that either

(i) $\lambda(\mathbf{v}) > \lambda(\mathbf{w})$,
(ii) $\lambda(\mathbf{v}) < \lambda(\mathbf{w})$ and $\mathbf{v} \in W_{\tau\alpha}$, or else
(iii) $\mathbf{v} = \langle \mathbf{\Phi}_\xi(\mathbf{v}_0\mathbf{v}_1 \cdots \mathbf{v}_\eta \cdots)\rangle_{\eta < \tau(\xi)}$ for some $\xi < \rho$, and $\mathbf{v}_\eta \in W_{\tau\alpha}$ with $\lambda(\mathbf{v}_\eta) < \lambda(\mathbf{w})$ for all $\eta < \tau(\xi)$. Since $\lambda(\mathbf{w}) > 1 = \lambda(\mathbf{x}_\zeta)$, it follows from (ii) that $\mathbf{x}_\zeta \in W_1$ for all $\zeta < \alpha$. Thus, W_1 satisfies 1.2(i). Suppose that $\mathbf{v}_\eta \in W_1$ for all $\eta < \tau(\xi)$, where $\xi < \rho$. We wish to show that $\mathbf{v} = \mathbf{\Phi}_\xi(\mathbf{v}_0\mathbf{v}_1 \cdots \mathbf{v}_\eta \cdots)$ $\in W_1$. If $\lambda(\mathbf{v}_\eta) < \lambda(\mathbf{w})$ for all η, then each $\mathbf{v}_\eta$ belongs to W_1 by virtue of conditions (ii) or (iii). In either case, $\mathbf{v}_\eta \in W_{\tau\alpha}$. Hence, $\mathbf{v} \in W_{\tau\alpha}$ by (iii). If $\lambda(\mathbf{v}_\eta) \geqq \lambda(\mathbf{w})$ for some η, then $\lambda(\mathbf{v}) \geqq \lambda(\mathbf{v}_\eta)) = \lambda(\mathbf{v}_\eta) + 1 > \lambda(\mathbf{w})$. Therefore, $\mathbf{v} \in W_1$ by (i). Thus, W_1 satisfies 1.2 (ii), so that $\mathbf{w} \in W_{\tau\alpha} \subseteq W_1$. However, $\mathbf{w}$ can belong to W_1 only by satisfying (iii), and this is the conclusion that was to be proved.

We now wish to prove that the representation of elements of $W_{\tau\alpha}$ described in Lemma 1.3 is unique. This uniqueness is an easy consequence of the following result.

1.4 LEMMA. If $\mathbf{w} \in W_{\tau\alpha}$ and $\mathbf{s}$ is a nonempty sequence of elements in $S_{\tau\alpha}$, then $\mathbf{ws} \notin W_{\tau\alpha}$.

Proof. Assuming the Lemma to be false, there exists $\mathbf{w} \in W_{\tau\alpha}$ and a nonempty $\mathbf{s}$ such that $\mathbf{ws} \in W_{\tau\alpha}$ and $\mathbf{w}$ is of minimal length among the words with this property. Note that $W_{\tau\alpha}$ does not contain the empty sequence because $W_{\tau\alpha} - \{\varnothing\}$ clearly satisfies conditions (i) and (ii) of Definition 1.2. Therefore, $\lambda(\mathbf{ws}) > 1$. By Lemma 1.3, $\mathbf{ws} = \langle \mathbf{\Phi}_\xi(\mathbf{w}_0\mathbf{w}_1 \cdots \mathbf{w}_\eta \cdots)\rangle_{\eta < \tau(\xi)}$, where

$\mathbf{w}_\eta \in W_{\tau\alpha}$. Thus, $\lambda(\mathbf{w}) > 1$, so that by Lemma 1.3 again, we have $\mathbf{w} = \langle \mathbf{\Phi}_\xi(\mathbf{v}_0\mathbf{v}_1 \cdots \mathbf{v}_\eta \cdots)\rangle_{\eta<\tau(\xi)}$, where $\mathbf{v}_\eta \in W_{\tau\alpha}$ and $\lambda(\mathbf{v}_\eta) < \lambda(\mathbf{w})$ for all $\eta < \tau(\xi)$. Thus, $\mathbf{w}_0\mathbf{w}_1 \cdots \mathbf{w}_\eta \cdots) = \mathbf{v}_0\mathbf{v}_1 \cdots \mathbf{v}_\eta \cdots)\mathbf{s}$. If $\mathbf{w}_\eta = \mathbf{v}_\eta$ for all $\eta < \tau(\xi)$, then $\mathbf{s}$ must be empty which is contrary to our hypothesis. Consequently, there is a minimal $\eta < \tau(\xi)$ such that $\mathbf{w}_\eta \neq \mathbf{v}_\eta$, and $\mathbf{w}_\eta \cdots) = \mathbf{v}_\eta \cdots)\mathbf{s}$. If $\lambda(\mathbf{w}_\eta) = \lambda(\mathbf{v}_\eta)$, then this last equality would imply that $\mathbf{w}_\eta = \mathbf{v}_\eta$. Thus, either $\lambda(\mathbf{v}_\eta) < \lambda(\mathbf{w}_\eta)$ or $\lambda(\mathbf{w}_\eta) < \lambda(\mathbf{v}_\eta)$, and we have $\mathbf{w}_\eta = \mathbf{v}_\eta\mathbf{s}'$, $\mathbf{s}' \neq \varnothing$, or $\mathbf{v}_\eta = \mathbf{w}_\eta\mathbf{s}''$, $\mathbf{s}'' \neq \varnothing$ in these respective cases. Since $\lambda(\mathbf{v}_\eta) < \lambda(\mathbf{w})$, both of these alternatives are contrary to the minimality of the length of $\mathbf{w}$. This contradiction proves the lemma.

1.5 Corollary. If $\langle \mathbf{\Phi}_\xi(\mathbf{v}_0\mathbf{v}_1 \cdots \mathbf{v}_\eta \cdots)\rangle_{\eta<\tau(\xi)} = \langle \mathbf{\Phi}_\xi(\mathbf{w}_0\mathbf{w}_1 \cdots \mathbf{w}_\eta \cdots)\rangle_{\eta<\tau(\xi)}$ where all $\mathbf{v}_\eta$ and $\mathbf{w}_\eta$ are in $W_{\tau\alpha}$, then $\mathbf{v}_\eta = \mathbf{w}_\eta$ for all $\eta < \tau(\xi)$.

Proof. Otherwise, there is a minimal η such that $\mathbf{v}_\eta \neq \mathbf{w}_\eta$. Then $\mathbf{v}_\eta \cdots) = \mathbf{w}_\eta \cdots)$, and either $\mathbf{w}_\eta = \mathbf{v}_\eta\mathbf{s}'$ or $\mathbf{v}_\eta = \mathbf{w}_\eta\mathbf{s}''$ for some nonempty $\mathbf{s}'$ and $\mathbf{s}''$. Either alternative contradicts Lemma 1.4.

We are now able to prove the main theorem of this section. An algebra of type τ is called *completely free* (on some generating set) if it is free in the class of all algebras of type τ.

1.6 Theorem. For $\xi < \rho$ and $\mathbf{w} \in (W_{\tau\alpha})^{\tau(\xi)}$, define

$$F_\xi(\mathbf{w}) = \langle \mathbf{\Phi}_\xi(\mathbf{w}(0)\mathbf{w}(1) \cdots \mathbf{w}(\eta) \cdots)\rangle_{\eta<\tau(\xi)}.$$

Then F_ξ is a $\tau(\xi)$-ary operation on $W_{\tau\alpha}$. The abstract algebra $\mathbf{W}_{\tau\alpha} = \langle W_{\tau\alpha}; F_\xi\rangle_{\xi<\rho}$ is completely free on the generating set $\{\mathbf{x}_\zeta \mid \zeta < \alpha\}$.

Proof. The conclusion that F_ξ is a $\tau(\xi)$-ary operation is a direct consequence of Definition 1.2 (ii). Moreover, the subalgebra of $\mathbf{W}_{\tau\alpha}$ generated by $\{\mathbf{x}_\zeta \mid \zeta < \alpha\}$ satisfies the conditions (i) and (ii) of 1.2, so that this subalgebra must be $W_{\tau\alpha}$. Hence $\{\mathbf{x}_\zeta \mid \zeta < \alpha\}$ generates $\mathbf{W}_{\tau\alpha}$. It remains to be proved that any mapping φ of $\{\mathbf{x}_\zeta \mid \zeta < \alpha\}$ to an algebra $\mathbf{A} = \langle A; G_\xi\rangle_{\xi<\rho}$ of type τ can be extended to a homomorphism of $\mathbf{W}_{\tau\alpha}$ to $\mathbf{A}$. By Zorn's lemma, there is an extension ψ of φ to a mapping of a set $T \subseteq W_{\tau\alpha}$ to A, such that ψ is maximal with the property

(i) $\quad \mathbf{w} \in (W_{\tau\alpha})^{\tau(\xi)}$ and $F_\xi(\mathbf{w}) \in T$ implies $\mathbf{w} \in T^{\tau(\xi)}$

and

$$\psi(F_\xi(\mathbf{w})) = G_\xi(\psi \circ \mathbf{w}).$$

If $T = W_{\tau\alpha}$, then ψ is a homomorphism, and we are finished. Suppose that $T \subset W_{\tau\alpha}$. Let $\mathbf{w}$ be of minimal length among the words of $W_\alpha - T$. Since $T \supseteq \{\mathbf{x}_\zeta \mid \zeta < \alpha\}$, it follows that $\lambda(\mathbf{w}) > 1$. Therefore, by Lemma 1.3, it is possible to find $\xi < \rho$ and $\mathbf{w}_\eta \in W_{\tau\alpha}$ such that $\lambda(\mathbf{w}_\eta) < \lambda(\mathbf{w})$ for all $\eta < \tau(\xi)$ and $\mathbf{w} = \mathbf{\Phi}_\xi(\mathbf{w}_0\mathbf{w}_1 \cdots \mathbf{w}_\eta \cdots)$. The minimality of $\lambda(\mathbf{w})$ implies that $\mathbf{w}_\eta \in T$ for all $\eta < \tau(\xi)$. Extend ψ to $S \cup \{\mathbf{w}\}$ by defining $\psi(\mathbf{w}) = G_\xi(\psi(\mathbf{w}_0), \psi(\mathbf{w}_1), \cdots, \psi(\mathbf{w}_\eta), \cdots)$. Using Corollary 1.5, it is not hard to see that the extended ψ satisfies (i). Thus, maximality is contradicted, so that T cannot be a proper subset of $W_{\tau\alpha}$.

The algebra $\mathbf{W}_{\tau\alpha}$ is called the *word algebra* of type τ on a standard alphabet of cardinality α.

2. Birkhoff's Theorem

Having defined words of type τ, it is now possible to say what we mean by an identity.

2.1 DEFINITION. An *identity* over the standard alphabet for algebras of type τ with α variables is a sequence of the form

$$\mathbf{v} = \mathbf{w},$$

where $\mathbf{v}$ and $\mathbf{w}$ are in $W_{\tau\alpha}$. An algebra $\mathbf{A}$ of type τ will be said to *satisfy the identity* $\mathbf{v} = \mathbf{w}$ if $\varphi(\mathbf{v}) = \varphi(\mathbf{w})$ for all $\varphi \in \mathrm{Hom}\,(\mathbf{W}_{\tau\alpha}, \mathbf{A})$.

The last part of this definition merits some comment. To say, for example, that some group $\mathbf{G}$ satisfies the identity $\mathbf{x}_0 \cdot \mathbf{x}_1 = \mathbf{x}_1 \cdot \mathbf{x}_0$ means that no matter what elements g_0 and g_1 of G are substituted for $\mathbf{x}_0$ and $\mathbf{x}_1$ in the words $\mathbf{x}_0 \cdot \mathbf{x}_1$ and $\mathbf{x}_1 \cdot \mathbf{x}_0$, the resulting elements $g_0 \cdot g_1$ and $g_1 \cdot g_0$ are the same. In other words, if φ is any mapping of $\{\mathbf{x}_0, \mathbf{x}_1\}$ to G, then the extension of φ to a homomorphism of the word algebra into $\mathbf{G}$ maps $\mathbf{x}_0 \cdot \mathbf{x}_1$ and $\mathbf{x}_1 \cdot \mathbf{x}_0$ onto the same elements. It is plain that this reasoning also applies to other familiar identities which occur in algebra. Indeed, the fact that the individual variables $\{\mathbf{x}_\zeta \mid \zeta < \alpha\}$ form a set of free generators of the word algebra $\mathbf{W}_{\tau\alpha}$ means that the homomorphisms of $\mathbf{W}_{\tau\alpha}$ to an algebra $\mathbf{A}$ of type τ coincide with the assignments of various values in A to these variables.

2.2 DEFINITION. Let α be a cardinal number. For each class $\mathfrak{A}$ of algebras of type τ, let $\mathfrak{E}_\alpha(\mathfrak{A})$ denote the set of all identities $\mathbf{v} = \mathbf{w}$ ($\mathbf{v}, \mathbf{w} \in W_{\tau\alpha}$) which are satisfied by every algebra in $\mathfrak{A}$. For each set $\mathfrak{E}$ of identities over a standard alphabet for algebras of type τ, let $\mathfrak{B}(\mathfrak{E})$ denote the class of all algebras of type τ that satisfy all of the identities in $\mathfrak{E}$. Any class of algebras of type τ that is of the form $\mathfrak{B}(\mathfrak{E})$ for some set $\mathfrak{E}$ of identities is called a *variety* or an *equationally definable class.*

2.3 LEMMA. Let α be a cardinal number. Let $\mathfrak{A}$ and $\mathfrak{A}'$ denote classes of algebras of type τ, and suppose that $\mathfrak{E}$ and $\mathfrak{E}'$ are sets of identities over a standard alphabet for algebras of type τ in α variables. Then

(a) $\mathfrak{A} \subseteq \mathfrak{A}'$ implies $\mathfrak{E}_\alpha(\mathfrak{A}') \subseteq \mathfrak{E}_\alpha(\mathfrak{A})$; $\mathfrak{E} \subseteq \mathfrak{E}'$ implies $\mathfrak{B}(\mathfrak{E}') \subseteq \mathfrak{B}(\mathfrak{E})$;
(b) $\mathfrak{A} \subseteq \mathfrak{B}(\mathfrak{E}_\alpha(\mathfrak{A}))$; $\mathfrak{E} \subseteq \mathfrak{E}_\alpha(\mathfrak{B}(\mathfrak{E}))$;
(c) $\mathfrak{B}(\mathfrak{E}_\alpha(\mathfrak{B}(\mathfrak{E}))) = \mathfrak{B}(\mathfrak{E})$; $\mathfrak{E}_\alpha(\mathfrak{B}(\mathfrak{E}_\alpha(\mathfrak{A}))) = \mathfrak{E}_\alpha(\mathfrak{A})$.

Proof. The statements under (a) and (b) are direct consequences of Definition 2.2, while the equalities in (c) are easily obtained from (a) and (b). From the inclusion $\mathfrak{E} \subseteq \mathfrak{E}_\alpha(\mathfrak{V}(\mathfrak{E}))$, it follows by (a) that $\mathfrak{V}(\mathfrak{E}_\alpha(\mathfrak{V}(\mathfrak{E}))) \subseteq \mathfrak{V}(\mathfrak{E})$; on the other hand, we obtain the reverse inclusion from the first part of (b) with $\mathfrak{A}$ replaced by $\mathfrak{V}(\mathfrak{E})$.

2.4 LEMMA. Let α be any cardinal number. Let $\mathfrak{A}$ be a class of algebras of type τ, and suppose that $\mathfrak{E}$ is a set of identities over a standard alphabet for algebras of type τ with α variables. Then

(a) $\mathfrak{E}_\alpha(\mathscr{Q}\mathfrak{A}) = \mathfrak{E}_\alpha(\mathscr{S}\mathfrak{A}) = \mathfrak{E}_\alpha(\mathscr{P}\mathfrak{A}) = \mathfrak{E}_\alpha(\mathfrak{A})$, and
(b) $\mathscr{Q}\mathfrak{V}(\mathfrak{E}) = \mathscr{S}\mathfrak{V}(\mathfrak{E}) = \mathscr{P}\mathfrak{V}(\mathfrak{E}) = \mathfrak{V}(\mathfrak{E})$.

Proof. (a) Since $\mathfrak{A} \subseteq \mathscr{Q}(\mathfrak{A})$, it follows that $\mathfrak{E}_\alpha(\mathscr{Q}\mathfrak{A}) \subseteq \mathfrak{E}_\alpha(\mathfrak{A})$. Similarly, $\mathfrak{E}_\alpha(\mathscr{S}\mathfrak{A}) \subseteq \mathfrak{E}_\alpha(\mathfrak{A})$ and $\mathfrak{E}_\alpha(\mathscr{P}\mathfrak{A}) \subseteq \mathfrak{E}_\alpha(\mathfrak{A})$. Assume that $\mathbf{v} = \mathbf{w}$ is an identity which is satisfied by every algebra in $\mathfrak{A}$. Suppose that $\mathbf{B} \in \mathscr{Q}\mathfrak{A}$, and ψ is an epimorphism of $\mathbf{A}$ to $\mathbf{B}$, where $\mathbf{A} \in \mathfrak{A}$. Let $\varphi \in \operatorname{Hom}(\mathbf{W}_{\tau\alpha}, \mathbf{B})$. Since $\psi(A) = B$, there is a mapping χ of $\{\mathbf{x}_\zeta \mid \zeta < \alpha\}$ into A satisfying $\psi(\chi(\mathbf{x}_\zeta)) = \varphi(\mathbf{x}_\zeta)$ for all $\zeta < \alpha$. In view of the fact proved in Theorem 1.6 that $\mathbf{W}_{\tau\alpha}$ is freely generated by $\{\mathbf{x}_\zeta \mid \zeta < \alpha\}$, χ extends to a homomorphism of $\mathbf{W}_{\tau\alpha}$ to $\mathbf{A}$. By Lemma 1.4.12, $\psi \circ \chi = \varphi$. Therefore, since $\mathbf{A}$ satisfies $\mathbf{v} = \mathbf{w}$, it follows that $\varphi(\mathbf{v}) = \psi(\chi(\mathbf{v})) = \psi(\chi(\mathbf{w})) = \varphi(\mathbf{w})$. This shows that $\mathbf{B}$ also satisfies $\mathbf{v} = \mathbf{w}$. Since $\mathbf{B} \in \mathscr{Q}\mathfrak{A}$ was arbitrary, we conclude that $\mathfrak{E}_\alpha(\mathscr{Q}\mathfrak{A}) = \mathfrak{E}_\alpha(\mathfrak{A})$. It is evident that if an identity is satisfied by an algebra $\mathbf{A}$, then it is satisfied by every subalgebra of $\mathbf{A}$. Hence, $\mathfrak{E}_\alpha(\mathscr{S}\mathfrak{A}) = \mathfrak{E}_\alpha(\mathfrak{A})$. Finally, suppose that $\{\mathbf{A}_i \mid i \in I\} \subseteq \mathfrak{A}$ and $\mathbf{C} = \Pi_{i \in I}\mathbf{A}_i$. Let $\varphi \in \operatorname{Hom}(\mathbf{W}_{\tau\alpha}, \mathbf{C})$. If π_i denotes the natural projection of $\mathbf{C}$ onto $\mathbf{A}_i$, then $\pi_i \circ \varphi \in \operatorname{Hom}(\mathbf{W}_{\tau\alpha}, \mathbf{A}_i)$. Consequently, $\pi_i(\varphi(\mathbf{v})) = \pi_i(\varphi(\mathbf{w}))$ for all $i \in I$. By Lemma 1.5.2, $\varphi(\mathbf{v}) = \varphi(\mathbf{w})$. This proves that $\mathfrak{E}_\alpha(\mathscr{P}\mathfrak{A}) = \mathfrak{E}_\alpha(\mathfrak{A})$.

(b) By (a) and Lemma 2.3, $\mathfrak{V}(\mathfrak{E}) = \mathfrak{V}(\mathfrak{E}_\alpha(\mathfrak{V}(\mathfrak{E}))) = \mathfrak{V}(\mathfrak{E}_\alpha(\mathscr{Q}\mathfrak{V}(\mathfrak{E}))) \supseteq \mathscr{Q}\mathfrak{V}(\mathfrak{E}) \supseteq \mathfrak{V}(\mathfrak{E})$. Similarly, $\mathfrak{V}(\mathfrak{E}) = \mathscr{S}\mathfrak{V}(\mathfrak{E})$ and $\mathfrak{V}(\mathfrak{E}) = \mathscr{P}\mathfrak{V}(\mathfrak{E})$.

Birkhoff's theorem asserts that the equalities of Lemma 2.4(b) characterize varieties.

2.5 THEOREM. A class $\mathfrak{A}$ of algebras of type τ is a variety if and only if $\mathscr{Q}\mathfrak{A} = \mathscr{S}\mathfrak{A} = \mathscr{P}\mathfrak{A} = \mathfrak{A}$.

The proof of this theorem occupies the remainder of the section. It involves two preliminary steps. First we show that there is a cardinal number α such that $\mathfrak{V}(\mathfrak{E}_\beta(\mathfrak{A})) = \mathfrak{V}(\mathfrak{E}_\alpha(\mathfrak{A}))$ for all $\beta \geqq \alpha$. Next it is shown that if $\mathbf{F}$ is a free $\mathfrak{A}$-algebra with β generators, then the homomorphism ψ of $\mathbf{W}_{\tau\beta}$ to $\mathbf{F}$ that maps $\{\mathbf{x}_\zeta \mid \zeta < \beta\}$ one-to-one onto a set of free generators of $\mathbf{F}$ satisfies $\Gamma_\psi = \{\langle \mathbf{v}, \mathbf{w} \rangle \mid \mathbf{v} = \mathbf{w} \text{ is in } \mathfrak{E}_\beta(\mathfrak{A})\}$. The main part of the proof of 2.5 is then completed by noting that any homomorphism ϑ of $\mathbf{W}_{\tau\beta}$ to an algebra $\mathbf{A}$ in $\mathfrak{V}(\mathfrak{E}_\beta(\mathfrak{A}))$ satisfies $\Gamma_\vartheta \supseteq \Gamma_\psi$, so that if ϑ is an epimorphism, it follows from Proposition 1.3.7 that $\mathbf{A} \in \mathscr{Q}\{\mathbf{F}\} \subseteq \mathfrak{A}$.

We proceed to fill in the details of this sketch.

2.6 LEMMA. Let $\mathbf{A} = \langle A; F_\xi \rangle_{\xi<\rho}$ be an algebra of type $\tau \in (\mathrm{Ord})^\rho$. Suppose that $X \subseteq A$ and $[X] = A$ in $\Sigma(\mathbf{A})$. Let α be an infinite cardinal number such that $|\tau(\xi)| \leqq \alpha$ for all $\xi < \rho$. Then

$$A = \bigcup\{[Y] \mid Y \subseteq X \quad \text{and} \quad |Y| \leqq \alpha\}.$$

Proof. Let $B = \bigcup\{[Y] \mid Y \subseteq X, |Y| \leqq \alpha\}$. Since $X \subseteq B$ and $[X] = A$, it is sufficient to prove that $B \in \Sigma(\mathbf{A})$. Let $\mathbf{a} \in B^{\tau(\xi)}$. Then for each $\eta < \tau(\xi)$, there exists $Y_\eta \subseteq X$ with $|Y_\eta| \leqq \alpha$ such that $\mathbf{a}(\eta) \in [Y_\eta]$. Let $Y = \bigcup_{\eta<\tau(\xi)} Y_\eta$. Then $|Y| \leqq |\tau(\xi)| \cdot \alpha \leqq \alpha^2 = \alpha$, and $[Y] \supseteq \bigcup_{\eta<\tau(\xi)}[Y_\eta]$. Thus, $\mathbf{a} \in [Y]^{\tau(\xi)}$, and $F_\xi(\mathbf{a}) \in [Y] \subseteq B$. Therefore, $B \in \Sigma(\mathbf{A})$.

2.7 LEMMA. Let $\mathfrak{C}$ be a class of abstract algebras of type τ. Suppose that $\mathbf{F}$ and $\mathbf{G}$ are free $\mathfrak{C}$-algebras on the respective generating sets X and Y. Assume that $0 < |X| \leqq |Y|$. Then for any set $Z \subseteq Y$ with $|Z| = |X|$, there exist $\psi \in \mathrm{Hom}\,(\mathbf{F}, \mathbf{G})$ and $\chi \in \mathrm{Hom}\,(\mathbf{G}, \mathbf{F})$ such that $\chi \circ \psi = I_F$ and $\psi(F) = [Z]$.

Proof. Let ψ map X one-to-one onto Z. Select an arbitrary $a \in F$, and define $\chi(y) = \psi^{-1}(y)$ for $y \in Z$ and $\chi(y) = a$ for $y \in Y - Z$. (Note that $F \neq \varnothing$, because $|X| > 0$.) Since $\mathbf{F}$ and $\mathbf{G}$ are free, it is possible to extend ψ and χ to homomorphisms. By Lemma 1.4.12 $\chi \circ \psi = I_F$. By Corollary 1.4.11, $\psi(F) = \psi([X]) = [Z]$.

We can now prove a result which is a key step in the proof of Birkhoff's Theorem and an interesting theorem in its own right.

2.8 PROPOSITION. Let α be an infinite cardinal number such that $\alpha \geqq |\tau(\xi)|$ for all $\xi < \rho$. Then for any class $\mathfrak{A}$ of algebras of type τ, and for each $\beta \geqq \alpha$,

$$\mathfrak{B}(\mathfrak{E}_\beta(\mathfrak{A})) = \mathfrak{B}(\mathfrak{E}_\alpha(\mathfrak{A})).$$

Proof. First note that $\alpha \leqq \beta$ implies $\mathfrak{E}_\alpha(\mathfrak{A}) \subseteq \mathfrak{E}_\beta(\mathfrak{A})$. Indeed, if $\mathbf{v} = \mathbf{w}$ is in $\mathfrak{E}_\alpha(\mathfrak{A})$, $\mathbf{A} \in \mathfrak{A}$, and $\varphi \in \mathrm{Hom}\,(\mathbf{W}_{\tau\beta}, \mathbf{A})$, we have $\varphi \restriction W_{\tau\alpha} \in \mathrm{Hom}\,(\mathbf{W}_{\tau\alpha}, \mathbf{A})$, so that $\varphi(\mathbf{v}) = \varphi(\mathbf{w})$. Hence, $\mathbf{v} = \mathbf{w}$ is also in $\mathfrak{E}_\beta(\mathfrak{A})$. It follows from Lemma 2.3 that $\mathfrak{B}(\mathfrak{E}_\alpha(\mathfrak{A})) \supseteq \mathfrak{B}(\mathfrak{E}_\beta(\mathfrak{A}))$. In order to prove that this inclusion is equality, suppose that $\mathbf{A}$ is an algebra of type τ with $\mathbf{A} \notin \mathfrak{B}(\mathfrak{E}_\beta(\mathfrak{A}))$. Then there is an identity $\mathbf{v} = \mathbf{w}$ in $\mathfrak{E}_\beta(\mathfrak{A})$, and a homomorphism $\varphi \in \mathrm{Hom}\,(\mathbf{W}_{\tau\beta}, \mathbf{A})$ such that $\varphi(\mathbf{v}) \neq \varphi(\mathbf{w})$. By Lemma 2.6, a set $Z \subseteq \{\mathbf{x}_\zeta \mid \zeta < \beta\}$ exists with $\mathbf{v} \in [Z]$, $\mathbf{w} \in [Z]$, and $|Z| = \alpha$. By Theorem 1.6 and Lemma 2.7, we can find $\psi \in \mathrm{Hom}\,(\mathbf{W}_{\tau\alpha}, \mathbf{W}_{\tau\beta})$ and $\chi \in \mathrm{Hom}\,(\mathbf{W}_{\tau\beta}, \mathbf{W}_{\tau\alpha})$ such that $\chi \circ \psi$ is the identity on $W_{\tau\alpha}$ and $\psi(W_{\tau\alpha}) = [Z]$. Let $\mathbf{v}_0$ and $\mathbf{w}_0$ in $W_{\tau\alpha}$ satisfy $\mathbf{v} = \psi(\mathbf{v}_0)$, $\mathbf{w} = \psi(\mathbf{w}_0)$. Then $\varphi \circ \psi \in \mathrm{Hom}\,(\mathbf{W}_{\tau\alpha}, \mathbf{A})$ and $(\varphi \circ \psi)(\mathbf{v}_0) = \varphi(\mathbf{v}) \neq \varphi(\mathbf{w}) = (\varphi \circ \psi)(\mathbf{w}_0)$. On the other hand, $\mathbf{v}_0 = \mathbf{w}_0$ is in $\mathfrak{E}_\alpha(\mathfrak{A})$. Indeed, suppose that $\mathbf{B} \in \mathfrak{A}$ and $\vartheta \in \mathrm{Hom}\,(\mathbf{W}_{\tau\alpha}, \mathbf{B})$. Then $\vartheta \circ \chi \in \mathrm{Hom}\,(\mathbf{W}_{\tau\beta}, \mathbf{B})$, so that $\vartheta(\mathbf{v}_0) = (\vartheta \circ \chi)(\mathbf{v}) = (\vartheta \circ \chi)(\mathbf{w}) = \vartheta(\mathbf{w}_0)$, the middle equality being a consequence of the fact that $\mathbf{v} = \mathbf{w}$ is in $\mathfrak{E}_\beta(\mathfrak{A})$. We conclude that $\mathbf{A}$ is not in $\mathfrak{B}(\mathfrak{E}_\alpha(\mathfrak{A}))$. Hence $\mathfrak{B}(\mathfrak{E}_\alpha(\mathfrak{A})) = \mathfrak{B}(\mathfrak{E}_\beta(\mathfrak{A}))$.

One other result is needed for the proof of the main theorem. For future reference, we present this lemma in a more general form than the proof of Birkhoff's theorem requires.

2.9 LEMMA. Let $\mathfrak{A}$ be a class of algebras of type τ. Suppose that $\mathbf{F}$ is an algebra of type τ that is generated by a set X of cardinality α. Let ψ be a homomorphism of $\mathbf{W}_{\tau\alpha}$ to $\mathbf{F}$ such that ψ maps $\{\mathbf{x}_\zeta \mid \zeta < \alpha\}$ one-to-one onto X. Define $\Delta = \{\langle \mathbf{v}, \mathbf{w}\rangle \in W_{\tau\alpha}^2 \mid \mathbf{v} = \mathbf{w} \text{ is in } \mathfrak{E}_\alpha(\mathfrak{A})\}$.

(a) If $\mathbf{F} \in \mathfrak{A}$, then $\Delta \subseteq \Gamma_\psi$.

(b) If every mapping of X to an algebra $\mathbf{A} \in \mathfrak{A}$ can be extended to a homomorphism of $\mathbf{F}$ to $\mathbf{A}$, then $\Gamma_\psi \subseteq \Delta$.

Proof. (a) If $\mathbf{F} \in \mathfrak{A}$ and $\mathbf{v} = \mathbf{w}$ is in $\mathfrak{E}_\alpha(\mathfrak{A})$, then $\mathbf{F}$ satisfies $\mathbf{v} = \mathbf{w}$, so that in particular $\psi(\mathbf{v}) = \psi(\mathbf{w})$.

(b) Suppose that $\psi(\mathbf{v}) = \psi(\mathbf{w})$. Let $\varphi \in \operatorname{Hom}(\mathbf{W}_{\tau\alpha}, \mathbf{A})$, where $\mathbf{A} \in \mathfrak{A}$. By assumption, the mapping $\varphi \circ (\psi \restriction \{\mathbf{x}_\zeta \mid \zeta < \alpha\})^{-1}$ has an extension to a homomorphism ϑ of $\mathbf{F}$ to $\mathbf{A}$. Then $\vartheta \circ \psi \in \operatorname{Hom}(\mathbf{W}_{\tau\alpha}, \mathbf{A})$, and $\vartheta\ (\psi(\mathbf{x}_\zeta)) = \varphi(\mathbf{x}_\zeta)$ for all $\zeta < \alpha$. By Lemma 1.4.12, $\vartheta \circ \psi = \varphi$. Hence, $\varphi(\mathbf{v}) = \vartheta(\psi(\mathbf{v})) = \vartheta(\psi(\mathbf{w})) = \varphi(\mathbf{w})$. Since $\mathbf{A}$ and φ were arbitrary, $\mathbf{v} = \mathbf{w}$ is in $\mathfrak{E}_\alpha(\mathfrak{A})$.

Proof of 2.5. If $\mathfrak{A}$ is a variety, then $\mathscr{Q}\mathfrak{A} = \mathscr{S}\mathfrak{A} = \mathscr{P}\mathfrak{A} = \mathfrak{A}$ by 2.4. Conversely, suppose that $\mathscr{Q}\mathfrak{A} = \mathscr{S}\mathfrak{A} = \mathscr{P}\mathfrak{A} = \mathfrak{A}$. The equality $\mathscr{P}\mathfrak{A} = \mathfrak{A}$ implies that $\mathfrak{A}$ contains all one element algebras of type τ. If $\mathfrak{A}$ contains no algebra with more than one element, then it is clear that $\mathfrak{A} = \mathfrak{V}(\mathbf{x}_0 = \mathbf{x}_1)$. Hence, we can assume that there are algebras in $\mathfrak{A}$ with more than one element. Then, since $\mathscr{S}\mathfrak{A} = \mathscr{P}\mathfrak{A} = \mathfrak{A}$, it follows from* Theorem 4.1.11 that any set X can be embedded as a set of free generators in a free $\mathfrak{A}$-algebra. Let α be the least infinite cardinal number satisfying $|\tau(\xi)| \leqq \alpha$ for all $\xi < \rho$. By 2.3, $\mathfrak{V}(\mathfrak{E}_\alpha(\mathfrak{A})) \supseteq \mathfrak{A}$. Suppose that $\mathbf{A} \in \mathfrak{V}(\mathfrak{E}_\alpha(\mathfrak{A}))$. To complete the proof of Theorem 2.5, we will show that $\mathbf{A} \in \mathfrak{A}$. Let β be a cardinal number such that $\beta \geqq \alpha$ and $\beta \geqq |A|$. By the choice of α, it follows from 2.8 that $\mathbf{A} \in \mathfrak{V}(\mathfrak{E}_\beta(\mathfrak{A}))$. Let $\mathbf{F}$ be a free $\mathfrak{A}$-algebra on the generating set $\{\mathbf{x}_\zeta \mid \zeta < \beta\}$. We have just noted that such a free algebra exists. Define $\psi \in \operatorname{Hom}(\mathbf{W}_{\tau\beta}, \mathbf{F})$ to be the homomorphism which extends the identity mapping on $\{\mathbf{x}_\zeta \mid \zeta < \beta\}$. Let ϑ be an epimorphism of $\mathbf{W}_{\tau\beta}$ to $\mathbf{A}$. The existence of ϑ is assured because $\beta \geqq |A|$ and $\mathbf{W}_{\tau\beta}$ is completely free. If $\langle \mathbf{v}, \mathbf{w}\rangle \in \Gamma_\psi$, then $\mathbf{v} = \mathbf{w}$ is an identity belonging to $\mathfrak{E}_\beta(\mathfrak{A})$ by Lemma 2.9. Therefore, $\vartheta(\mathbf{v}) = \vartheta(\mathbf{w})$ because $\mathbf{A} \in \mathfrak{V}(\mathfrak{E}_\beta(\mathfrak{A}))$ satisfies all of the identities in $\mathfrak{E}_\beta(\mathfrak{A})$. In other words, we have shown that $\Gamma_\psi \subseteq \Gamma_\vartheta$. By the homomorphism theorem (1.3.7), there is a

* Although the existence theorem for free $\mathfrak{A}$-algebras is an essential part of the proof of 2.5, it is possible to obtain the special case of Theorem 4.1.11, which we need, directly from Theorem 1.6. In fact, if $\mathfrak{A}$ is a class of abstract algebras of type τ such that $\mathscr{Q}\mathfrak{A} = \mathscr{S}\mathfrak{A} = \mathscr{P}\mathfrak{A} = \mathfrak{A}$, and $\mathfrak{A}$ contains an algebra with more than one element, then it is easy to see that $\mathbf{W}_{\tau\alpha}/N_{\mathfrak{A}}(\mathbf{W}_{\tau\alpha})$ is a free $\mathfrak{A}$-algebra with α generators (see Problem **12**, Chapter 1).

homomorphism φ of $\mathbf{F}$ to $\mathbf{A}$ satisfying $\vartheta = \varphi \circ \psi$. Since ϑ is onto, φ is an epimorphism. Consequently, $\mathbf{A} \in \mathscr{Q}\{\mathbf{F}\} \subseteq \mathscr{Q}\mathfrak{A} = \mathfrak{A}$.

3. The Generation of Varieties

If $\mathfrak{A}$ is a class of algebras of type τ, then by Birkhoff's theorem and Corollary 1.6.4, the class

$$\hat{\mathfrak{A}} = \mathscr{Q}\mathscr{S}\mathscr{P}\mathfrak{A}$$

is a variety. Evidently, $\hat{\mathfrak{A}}$ is the smallest variety containing $\mathfrak{A}$, so that we will call it the variety generated by $\mathfrak{A}$. Our objective in this final section is to obtain some alternative descriptions of $\hat{\mathfrak{A}}$. The first characterization is in terms of identities. It depends on a cardinal number invariant of classes of algebras.

3.1 Definition. Let $\mathfrak{A}$ be a class of abstract algebras of type τ. The *axiom rank* of $\mathfrak{A}$ is the cardinal number

$$r(\mathfrak{A}) = \min \{\alpha \in \text{Card} \mid \mathfrak{V}(\mathfrak{E}_\beta(\mathfrak{A})) = \mathfrak{V}(\mathfrak{E}_\alpha(\mathfrak{A})) \text{ for all } \beta \geqq \alpha\}.$$

By Proposition 2.8, $r(\mathfrak{A})$ is well defined. Moreover, the following facts are easily obtained from the work of the preceding section.

3.2 Lemma. Let $\mathfrak{A}$ be a class of algebras of type $\tau \in \text{Ord}^\rho$.

(a) $r(\mathfrak{A}) \leqq \sup (\{|\tau(\xi)| \mid \xi < \rho\} \cup \{\aleph_0\})$.

(b) $r(\mathfrak{A}) = r(\hat{\mathfrak{A}})$.

(c) If $\mathfrak{A} = \mathfrak{V}(\mathfrak{E})$, where $\mathfrak{E}$ is a set of identities in the words of $W_{\tau\alpha}$, then $r(\mathfrak{A}) \leqq \alpha$.

Proof. (a) follows from 2.8. By Lemma 2.4(a), $\mathfrak{E}_\alpha(\mathfrak{A}) = \mathfrak{E}_\alpha(\hat{\mathfrak{A}})$ for all cardinal numbers α. The equality in (b) follows from this observation and the definitions of $r(\mathfrak{A})$ and $r(\hat{\mathfrak{A}})$. In order to prove (c), note that if $\beta \geqq \alpha$, then $\mathfrak{E}_\beta(\mathfrak{A}) \supseteq \mathfrak{E}_\alpha(\mathfrak{A}) \supseteq \mathfrak{E}$. Hence, $\mathfrak{A} = \mathfrak{V}(\mathfrak{E}) \supseteq \mathfrak{V}(\mathfrak{E}_\alpha(\mathfrak{A})) \supseteq \mathfrak{V}(\mathfrak{E}_\beta(\mathfrak{A})) \supseteq \mathfrak{A}$ by Lemma 2.3.

Our characterization of the variety generated by a class of algebras is an easy consequence of the results of section 2.

3.3 Proposition. Let $\mathfrak{A}$ be a class of similar algebras that contains a nonempty algebra. If $\alpha \geqq r(\mathfrak{A})$, then

$$\hat{\mathfrak{A}} = \mathfrak{V}(\mathfrak{E}_\alpha(\mathfrak{A})).$$

Proof. By Birkhoff's theorem 2.5, there is a cardinal number β and a set $\mathfrak{E}$ of identities in the words of $W_{\tau\beta}$ such that $\hat{\mathfrak{A}} = \mathfrak{V}(\mathfrak{E})$. By Lemma 3.2, $r(\mathfrak{A}) = r(\hat{\mathfrak{A}}) \leqq \beta$. Since $\mathfrak{E} \subseteq \mathfrak{E}_\beta(\mathfrak{V}(\mathfrak{E})) = \mathfrak{E}_\beta(\hat{\mathfrak{A}})$, it follows that $\hat{\mathfrak{A}} \subseteq \mathfrak{V}(\mathfrak{E}_\alpha(\hat{\mathfrak{A}})) = \mathfrak{V}(\mathfrak{E}_\beta(\hat{\mathfrak{A}})) \subseteq \mathfrak{V}(\mathfrak{E}) = \hat{\mathfrak{A}}$.

It follows from 3.2(c) and 3.3 that if $\mathfrak{A}$ is a variety, then $r(\mathfrak{A})$ is the smallest cardinal number α such that $\mathfrak{A}$ is definable by a set of identities in α variables.

It is convenient to designate by $\hat{\mathbf{A}}$ the variety which is generated by the one element class $\{\mathbf{A}\}$ (instead of writing $\{\hat{\mathbf{A}}\}$). An algebra $\mathbf{A}$ is called *generic for the variety* $\mathfrak{A}$ if $\hat{\mathbf{A}} = \mathfrak{A}$. One might expect that the varieties that contain a generic algebra would be special, but this is not the case.

3.4 PROPOSITION. Let $\mathfrak{A}$ be a variety of algebras of type $\tau \in (\mathrm{Ord})^\rho$ and suppose that $\mathfrak{A}$ contains an algebra with more than one element. Let α be an infinite cardinal number that satisfies $\alpha \geqq |\tau(\xi)|$ for all $\xi < \rho$. Then the free $\mathfrak{A}$-algebra with α generators is generic for $\mathfrak{A}$.

Proof. Let $\mathbf{F}$ be a free $\mathfrak{A}$-algebra with α generators. The existence of such an algebra is guaranteed by 4.1.11. By 2.9, $\mathfrak{E}_\alpha(\{\mathbf{F}\}) \subseteq \mathfrak{E}_\alpha(\mathfrak{A})$. On the other hand, $\mathbf{F} \in \mathfrak{A}$, so that $\mathfrak{E}_\alpha(\mathfrak{A}) \subseteq \mathfrak{E}_\alpha(\{\mathbf{F}\})$. By 3.2, $\alpha \geqq r(\mathfrak{A})$, and $\alpha \geqq r(\hat{\mathbf{F}})$. Therefore, $\hat{\mathbf{F}} = \mathfrak{B}(\mathfrak{E}_\alpha(\{\mathbf{F}\}) = \mathfrak{B}(\mathfrak{E}_\alpha(\mathfrak{A})) = \mathfrak{A}$.

3.5 EXAMPLE. Let $\mathfrak{A}$ be the class of all Abelian groups. By definition, $\mathfrak{A}$ is a variety. An Abelian group $\mathbf{A}$ is said to have *exponent* n if $nx = 0$ for all $x \in A$, and n is the least natural number with this property. If there is no n such that $nx = 0$ for all x, then we will say that $\mathbf{A}$ has infinite exponent. For any natural number n, let $\mathfrak{A}_n$ denote the class of all Abelian groups that satisfy the identity $\mathbf{x}_0 + (\mathbf{x}_0 + (\mathbf{x}_0 + \cdots + \mathbf{x}_0) \cdots) = \mathbf{0}$, with exactly n occurrences of $\mathbf{x}_0$ on the left-hand side. It is obvious that $\mathfrak{A}_n$ contains every Abelian group that has exponent n. The main result of this example is a converse statement.

3.6 LEMMA. (a) If $\mathbf{A} \in \mathfrak{A}$ has exponent n, then $\hat{\mathbf{A}} = \mathfrak{A}_n$.

(b) If $\mathbf{A} \in \mathfrak{A}$ has infinite exponent, then $\hat{\mathbf{A}} = \mathfrak{A}$.

The proofs of both of these statements make use of an elementary fact: if $x \in A$ satisfies $nx = 0$, then the order of x divides n. Indeed, suppose that m is the smallest natural number satisfying $mx = 0$. The m elements $0, x, 2x, \cdots, (m-1)x$ are distinct, since $ix = jy$, $0 \leqq i < j < m$, implies $(j-i)x = 0$. Moreover, any integer k can be expressed in the form $k = qm + r$, where q and r are integers and $0 \leqq r < m$, so that $kx = q(mx) + rx = rx$. Thus, $m = 0(x)$, and $nx = 0$ if and only if m divides n.

Suppose that $\mathbf{A} \in \mathfrak{A}$ has finite exponent n. The elementary result that we have just established implies that n is the least common multiple of the orders of the elements of $\mathbf{A}$. Thus, there is a finite set $\{a_0, a_1, \cdots, a_{t-1}\} \subseteq A$ such that n is the least common multiple of $\{0(a_0), 0(a_1), \cdots, 0(a_{t-1})\}$. Let $\mathbf{a} \in A^t$ be defined by $\mathbf{a}(i) = a_i$. Clearly, $k\mathbf{a} = 0$ if and only if $ka_i = 0$ for all i. Therefore $0(\mathbf{a}) = n$. This argument shows that $\hat{\mathbf{A}}$ contains a cyclic group of order n, namely $\mathbf{C} = \mathbf{A}^t \restriction [\mathbf{a}]$. We now observe that $\mathbf{C}$ is a free $\mathfrak{A}_n$-algebra with the free generator $\mathbf{a}$. In fact, if $\mathbf{B} \in \mathfrak{A}_n$, and $b \in B$, then $0(b)$ divides n. It follows easily that the mapping $k\mathbf{a} \to kb$ is a well-defined homomorphism of **C** to **B**. Since the free $\mathfrak{A}_n$-algebra with any (cardinal) number of generators is a direct sum of free $\mathfrak{A}_n$-algebras with one generator (see Examples 4.2.2 and 4.2.3), and the direct sum is a subalgebra of the direct product, we conclude from 3.2 that $\mathfrak{A}_n = \hat{\mathbf{A}}$.

The proof of statement (b) follows a similar pattern. First notice that the orders of the elements of **A** form an unbounded set (or include ∞). Therefore, an infinite direct product of copies of **A** contains an infinite cyclic group, that is a free $\mathfrak{A}$-algebra with one generator. Thus, $\hat{\mathbf{A}} = \mathfrak{A}$ as before.

As a consequence of this lemma, it follows that every proper subvariety of the variety of all Abelian groups is of the form $\mathfrak{A}_n$ for some natural number n.

We conclude by deriving a sufficient condition for an algebra to belong to a variety. This result will then be used to prove a basic theorem on the representation of countably complete Boolean algebras.

3.7 PROPOSITION. Let $\mathfrak{A}$ be a variety of algebras of type τ. Define α to be the least infinite cardinal number satisfying $|\tau(\xi)| \leqq \alpha$ for all $\xi < \rho$. Let **A** be an algebra of type τ. Suppose that for every subsystem† $\mathbf{B} = \langle B; G_\xi \rangle_{\xi<\rho}$ of **A** such that $|G_\xi| \leqq \alpha$, and for every $a \neq b$ in B, there exists a homomorphism φ of **B** to an algebra in $\mathfrak{A}$ such that $\varphi(a) \neq \varphi(b)$. Then $\mathbf{A} \in \mathfrak{A}$.

Proof. By 3.2 and 3.3, $\mathfrak{A} = \mathfrak{B}(\mathfrak{E}_\alpha(\mathfrak{A}))$. Hence, it suffices to prove that if an identity $\mathbf{v} = \mathbf{w}(\mathbf{v}, \mathbf{w}$ in $W_{\tau\alpha})$ is not satisfied in **A**, then there is an algebra of $\mathfrak{A}$ which does not satisfy the identity. It is convenient to introduce a technical definition. Let $\mathbf{w} \in W_{\tau\alpha}$ and suppose that **C** is a subsystem of $\mathbf{W}_{\tau\alpha}$ such that $\{\mathbf{x}_\zeta \mid \zeta < \alpha\} \subseteq C$. We will say that **w** is strongly contained in **C** if $\mathbf{w} \in C$, and when $\psi \in \mathrm{Hom}(\mathbf{C}, \mathbf{D})$, $\vartheta \in \mathrm{Hom}(\mathbf{W}_{\tau\alpha}, \mathbf{D})$ satisfy $\psi(\mathbf{x}_\zeta) = \vartheta(\mathbf{x}_\zeta)$ (**D** being any algebra of type τ), it follows that $\psi(\mathbf{w}) = \vartheta(\mathbf{w})$.

3.8 LEMMA. Let $\mathbf{w} \in W_{\tau\alpha}$. Then there is a subsystem $\mathbf{C} = \langle C; H_\xi \rangle_{\xi<\rho}$ of $\mathbf{W}_{\tau\alpha}$ such that $|H_\xi| \leqq \alpha$ for all $\xi < \rho$, $\{\mathbf{x}_\zeta \mid \zeta < \alpha\} \subseteq C$, and **w** is strongly contained in **C**.

To prove this lemma, let V be the set of all $\mathbf{w} \in W_{\tau\alpha}$ that are strongly contained in some **C**, satisfying the conditions of the lemma. If $C = \{\mathbf{x}_\zeta \mid \zeta < \alpha\}$ and $H_\xi = \varnothing$ for all $\xi < \rho$, then each $\mathbf{x}_\zeta$ is strongly contained in $\langle C; H_\xi \rangle_{\xi<\rho}$. Thus, $V \supseteq \{\mathbf{x}_\zeta \mid \zeta < \alpha\}$. The proof of the lemma will be completed by showing that $V \in \Sigma(\mathbf{W}_{\tau\alpha})$, since in this case $V \supseteq [\{\mathbf{x}_\zeta \mid \zeta < \alpha\}] = W_{\tau\alpha}$. Let $\{\mathbf{w}_\eta \mid \eta < \tau(\sigma)\} \subseteq V$, where $\sigma < \rho$. For each $\eta < \tau(\sigma)$, let $\mathbf{C}_\eta = \langle C_\eta; H_{\eta\xi} \rangle_{\xi<\rho}$ be a subsystem of $\mathbf{W}_{\tau\alpha}$ satisfying the conditions of the lemma such that $\mathbf{w}_\eta$ is strongly contained in $\mathbf{C}_\eta$. Define $C = (\bigcup_{\eta<\tau(\sigma)} C_\eta) \cup \{\mathbf{\Phi}_\xi(\mathbf{w}_0\mathbf{w}_1 \cdots)\}$, $H_\xi = \bigcup_{\eta<\tau(\sigma)} H_{\eta\xi}$ for $\xi \neq \sigma$, and

$$H_\sigma = (\bigcup_{\eta<\tau(\sigma)} H_{\eta\sigma}) \cup \{\langle \mathbf{w}_0\mathbf{w}_1 \cdots \mathbf{\Phi}_\xi(\mathbf{w}_0\mathbf{w}_1 \cdots)\rangle\}$$

† See Definition 1.4.1. Here, **A** is considered as a relational system.

Then $\mathbf{C} = \langle C; H_\xi \rangle_{\xi<\rho}$ is a subsystem of $\mathbf{W}_{\tau\alpha}$, which satisfies the conditions of the lemma and strongly contains $\mathbf{\Phi}_\xi(\mathbf{w}_0\mathbf{w}_1 \cdots)$. Therefore, $\mathbf{\Phi}_\xi(\mathbf{w}_0\mathbf{w}_1 \cdots) \in V$, and the proof of the lemma is finished.

Suppose now that $\mathbf{v}$ and $\mathbf{w}$ are words in $W_{\tau\alpha}$ such that $\mathbf{A}$ does not satisfy the identity $\mathbf{v} = \mathbf{w}$. Then there is a homomorphism χ of $\mathbf{W}_{\tau\alpha}$ to $\mathbf{A}$ such that $\chi(\mathbf{v}) \neq \chi(\mathbf{w})$. Let $\mathbf{C} = \langle C; H_\xi \rangle_{\xi<\rho}$ and $\mathbf{D} = \langle D; K_\xi \rangle_{\xi<\rho}$ be subsystems of $\mathbf{W}_{\tau\alpha}$ that satisfy the conditions of the lemma, and strongly contain $\mathbf{v}$ and $\mathbf{w}$ respectively. Define $B = \chi(C) \cup \chi(D)$, $G_\xi = \chi \circ H_\xi \cup \chi \circ K_\xi$. Then $\mathbf{B} = \langle B; G_\xi \rangle_{\xi<\rho}$ is a subsystem of $\mathbf{A}$, and $\chi(\mathbf{v})$ and $\chi(\mathbf{w})$ are distinct elements of B. Since $|G_\xi| \leqq |H_\xi \cup K_\xi| \leqq 2\alpha = \alpha$, it follows from the hypothesis of the proposition that there is a homomorphism φ of $\mathbf{B}$ to an algebra $\mathbf{E} \in \mathfrak{A}$ such that $\varphi(\chi(\mathbf{v})) \neq \varphi(\chi(\mathbf{w}))$. Let $\vartheta \in \operatorname{Hom}(\mathbf{W}_{\tau\alpha}, \mathbf{E})$ be determined by the condition $\vartheta(\mathbf{x}_\zeta) = \varphi(\chi(\mathbf{x}_\zeta))$ for all $\zeta < \alpha$. Since $\mathbf{v}$ and $\mathbf{w}$ are strongly contained in $\mathbf{C}$ and $\mathbf{D}$, respectively and $(\varphi \circ \chi) \restriction C \in \operatorname{Hom}(\mathbf{C}, \mathbf{E})$, $(\varphi \circ \chi) \restriction D \in \operatorname{Hom}(\mathbf{D}, \mathbf{E})$, it follows that $\vartheta(\mathbf{v}) = \varphi(\chi(\mathbf{v})) \neq \varphi(\chi(\mathbf{w})) = \vartheta(\mathbf{w})$. Consequently, $\mathbf{v} = \mathbf{w}$ is not in $\mathfrak{E}_\alpha(\mathfrak{A})$. This completes the proof of 3.7.

3.9 EXAMPLE. Let $\mathfrak{B}$ be the class of all ω-complete Boolean algebras (see Example 4.1.8). Let $\mathfrak{A} = \hat{\mathbf{2}}$, where $\mathbf{2}$ is the two-element Boolean algebra. The class $\mathfrak{A} = \hat{\mathbf{2}}$ can be characterized in a more interesting way. We have observed in Example 2.4.2 that a product of α copies of $\mathbf{2}$ is isomorphic (as a Boolean algebra) to the Boolean algebra of all subsets of α. It is easy to see that the ω-ary join and meet operations in the product correspond under this isomorphism to countable union and intersection. Thus, $\mathscr{S}\mathscr{P}\mathbf{2}$ consists of all ω-complete Boolean algebras that are isomorphic to σ-fields‡ of sets, that is, Boolean algebras of sets that are closed under countable unions and intersections. Finally, $\mathfrak{A} = \mathscr{Q}\mathscr{S}\mathscr{P}\mathbf{2}$ is the class of all σ-representable Boolean algebras, that is, Boolean algebras which are σ-homomorphic images (by a homomorphism that preserves countable joins and meets) of σ-fields of sets.

Since $\mathbf{2} \in \mathfrak{B}$ and $\mathfrak{B}$ is a variety, it follows that $\mathfrak{A} \subseteq \mathfrak{B}$. We wish to prove that $\mathfrak{A} = \mathfrak{B}$. In view of the above remarks, this equality is equivalent to the fundamental theorem of Loomis [35] and Sikorski [44] on the representation of σ-complete Boolean algebras.

3.10 THEOREM. Every countably complete Boolean algebra is σ-representable.

Proof. We will use Proposition 3.7 to show that $\mathfrak{B} \subseteq \mathfrak{A}$. Let $\mathbf{B} \in \mathfrak{B}$. Let $\{a_{mn} \mid m < \omega, n < \omega\} \subseteq B$ and $\{b_{mn} \mid m < \omega, n < \omega\} \subseteq B$. Suppose that $a_m = \bigvee_{n<\omega} a_{mn}$, $b_m = \bigwedge_{n<\omega} b_{mn}$. We will prove that for each pair of distinct elements a and b in B there is a mapping φ of B to 2 such that φ is a Boolean

‡ It would be logical to call these algebras ω-fields. However, σ-field is the traditional terminology.

algebra homomorphism, $\varphi(a) \neq \varphi(b)$, and $\varphi(a_m) = \bigvee_{n<\omega}\varphi(a_{mn})$, $\varphi(b_m) = \bigwedge_{n<\omega}\varphi(b_{mn})$ for all $m < \omega$. Having established this, the desired conclusion that $\mathbf{B} \in \mathfrak{A}$ will follow from 3.7.

Define elements c_{mn} and c_m by $c_{2mn} = a_{mn}$, $c_{2m} = a_m$ and $c_{2m+1\ n} = b'_{nm}$, $c_{m+1} = b'_m$. Note that $\bigvee_{n<\omega}c_{mn} = c_m$ for all m. Indeed, $b_m = \text{g.l.b.}$ $\{b_{mn} \mid n < \omega\}$ implies $c_{2m+1} = b'_m \geqq b'_{mn} = c_{2m+1\ n}$ for all $n < \omega$, and if $c \geqq c_{2m+1\ n}$ for all n, then $c' \leqq c'_{2m+1\ n} = b_{mn}$, so that $c' \leqq b_m$, and $c \geqq c_{2m+1}$. Next, we observe that $(a \wedge b) \vee (a' \wedge b') \neq u$. Otherwise, $a = a \wedge u = a \wedge ((a \wedge b) \vee (a' \wedge b')) = (a \wedge a \wedge b) \vee (a \wedge a' \wedge b) = (a \wedge b) \vee (o \wedge b) = (a \wedge b) \vee o = a \wedge b$, and similarly, $b = a \wedge b$. However, this is contrary to $a \neq b$. By induction, define a sequence of elements $d_0, d_1, d_2, \cdots$ different from u, with $d_0 \leqq d_1 \leqq d_2 \leqq \cdots$ such that $d_0 = (a \wedge b) \vee (a' \wedge b')$, and for each $m < \omega$, either $c_m \leqq d_{m+1}$, or $c'_{mn} \leqq d_{m+1}$ for some n. Assume that $d_0, d_1, \cdots, d_m$ have been defined, where $m \geq 0$. If $d_m \vee c'_{mn} \neq u$ for some smallest n, let $d_{m+1} = d_m \vee c'_{mn}$. Otherwise, $c_{mn} \leqq d_m$ for all $n < \omega$, so that $c_m = \bigvee_{n<\omega}c_{mn} \leqq d_m$. Thus, we can take $d_{m+1} = d_m$ in this case.

Define Γ to be the set of all $\langle x, y\rangle \in B^2$ such that $x \vee d_n = y \vee d_n$ for some $n < \omega$. By an argument like the one given in Lemma 4.3.7, it follows that Γ is a congruence relation on $\langle B; \vee, \wedge, ', o, u\rangle$ (that is, $\mathbf{B}$ considered only as a Boolean algebra). Note that $\Gamma \neq B^2$. In fact $\langle o, u\rangle \in \Gamma$ because $o \vee d_m = d_m \neq u = u \vee d_m$ for all $m < \omega$. The proof of the theorem in Example 2.4.2 shows that there is a Boolean algebra homomorphism ψ of $\mathbf{B}/\Gamma$ to $\mathbf{2}$. Let $\varphi = \psi \circ \bar{\Gamma}$. Note that if $x \leqq d_m$ for some m, then $\langle o, x\rangle \in \Gamma$, so that $\varphi(x) = 0$. In particular, $(\varphi(a) \wedge \varphi(b)) \vee (\varphi(a)' \wedge \varphi(b)') = \varphi(d_0) = 0$, which implies that $\varphi(a) \neq \varphi(b)$. Moreover, for each $m < \omega$, either $\varphi(c_m) = 0$, or else there is an $n < \omega$ such that $\varphi(c_{mn})' = \varphi(c'_{mn}) = 0$ (so that $\varphi(c_{mn}) = u$). Consequently, $\varphi(c_m) = \bigvee_{n<\omega}\varphi(c_{mn})$. Letting m range over the even ordinals gives $\varphi(a_m) = \bigvee_{n<\omega}\varphi(a_{mn})$ for all $m < \omega$. The odd values of m give the conclusion that $\varphi(b_m)' = \varphi(b'_m) = \bigvee_{n<\omega}\varphi(b'_{mn})$. Consequently, $\varphi(b_m) = \varphi(b'_m)' = (\bigvee_{n<\omega}\varphi(b'_{mn}))' = \bigwedge_{n<\omega}\varphi(b'_{mn})' = \bigwedge_{n<\omega}\varphi(b_{mn})$.

Problems

1. (a) Let $\mathfrak{A}$ be a class of algebras of type τ satisfying $\mathscr{S}\mathfrak{A} = \mathfrak{A}$. Define $\Delta = \{\langle \mathbf{v}, \mathbf{w}\rangle \in W^2_{\tau\alpha} \mid \mathbf{v} = \mathbf{w} \text{ is in } \mathfrak{E}_\alpha(\mathfrak{A})\}$. Prove that $\Delta = N_{\mathfrak{A}}(\mathbf{W}_{\tau\alpha}) = \bigcap\{\Gamma \in \Theta(\mathbf{W}_{\tau\alpha}) \mid \mathbf{W}_{\tau\alpha}/\Gamma \in \mathfrak{A}\}$. Also show that Δ is invariant under endomorphisms of $\mathbf{W}_{\tau\alpha}$, that is, if $\varphi \in \text{Hom}\,(\mathbf{W}_{\tau\alpha}, \mathbf{W}_{\tau\alpha})$, then $\varphi \circ \Delta \subseteq \Delta$.

(b) Let $\mathfrak{E}$ be a set of identities in the words of type τ in α variables. Define $\Delta = \{\langle \mathbf{v}, \mathbf{w}\rangle \mid \mathbf{v} = \mathbf{w} \text{ is in } \mathfrak{E}\}$. Assume that Δ is a congruence relation on $\mathbf{W}_{\tau\alpha}$ that is invariant under all endomorphisms of $\mathbf{W}_{\tau\alpha}$. Define $\mathbf{F} = \mathbf{W}_{\tau\alpha}/\Delta$. Prove that $\mathfrak{E}_\alpha(\{\mathbf{F}\}) = \mathfrak{E}$.

2. (a) Let $\mathfrak{A}$ be a class of algebras of type $\tau \in \text{Ord}^\rho$. Use 3.4 to show that there is a *set* $\mathfrak{B} \subseteq \mathfrak{A}$ such that $\hat{\mathfrak{B}} = \hat{\mathfrak{A}}$.

(b) Suppose that α is an infinite cardinal number, such that $\tau \in \alpha^\rho$. Suppose that **A** is an algebra of type τ such that $|A| \leqq \alpha$. Assume that **v** and **w** are words in $W_{\tau\alpha}$ such that there is a homomorphism $\varphi \in \text{Hom}(\mathbf{W}_{\tau\alpha}, \mathbf{A})$, which satisfies $\varphi(\mathbf{v}) \neq \varphi(\mathbf{w})$. Prove that there is an epimorphism ψ of $\mathbf{W}_{\tau\alpha}$ to **A** such that $\psi(\mathbf{v}) \neq \psi(\mathbf{w})$.

(c) Let α satisfy the conditions of **(b)**. Suppose that $\mathfrak{B}$ is a set of algebras of type τ such that $|A| \leqq \alpha$ for all $\mathbf{A} \in \mathfrak{B}$. Prove that:

$$\bigcap\{\Gamma \in \Theta(\mathbf{W}_{\tau\alpha}) \mid \mathbf{W}_{\tau\alpha}/\Gamma \in \mathscr{I}\mathfrak{B}\} = \bigcap\{\Gamma \in \Theta(\mathbf{W}_{\tau\alpha}) \mid \mathbf{W}_{\tau\alpha}/\Gamma \in \mathscr{S}\mathfrak{B}\}.$$

(d) Use the results of **(a)** and **(c)** above and Problem **8** in Chapter 4 to prove that if $\mathfrak{A}$ is a class of algebras of type τ, then $\mathscr{Q}\mathscr{R}\mathfrak{A} = \hat{\mathfrak{A}}$.

3. Let τ be a finitary type (that is, $\tau \in \omega^\rho$).

(a) Prove that $\lambda(\mathbf{w}) < \omega$ for all $\mathbf{w} \in W_{\tau\alpha}$ (where α is any cardinal number).

(b) Prove that $W_{\tau\omega} = \bigcup_{n<\omega} W_{\tau n}$.

4. Let $\mathfrak{A}$ be a class of algebras of type τ. Let X be a nonempty generating set for an algebra $\mathbf{F} \in \hat{\mathfrak{A}}$. Assume that $|X| = \alpha$. Prove that the following conditions are equivalent:

(a) every mapping of X to $\mathbf{A} \in \mathfrak{A}$ can be extended to a homomorphism of **F** to **A**;

(b) **F** is a free $\hat{\mathfrak{A}}$-algebra on the generating set X;

(c) **F** is a free $\mathfrak{B}(\mathfrak{E}_\alpha(\mathfrak{A}))$-algebra on the generating set X. (*Hint:* by 4.1.11, X can be embedded as a set of free generators in a free $\hat{\mathfrak{A}}$-algebra **G**, and in a free $\mathfrak{B}(\mathfrak{E}_\alpha(\mathfrak{A}))$-algebra **H**. A one-to-one mapping of $\{\mathbf{x}_\zeta \mid \zeta < \alpha\}$ onto X extends to homomorphisms φ of $\mathbf{W}_{\tau\alpha}$ to **F**, ψ of $\mathbf{W}_{\tau\alpha}$ to **G**, and χ of $\mathbf{W}_{\tau\alpha}$ to **H**. Deduce from 2.9 that **(a)** implies $\Gamma_\varphi = \Gamma_\psi = \Gamma_\chi$.)

5. (a) Let **A** be an algebra of type τ. Suppose that α is any cardinal number. Define $\Phi = A^\alpha$, and $\mathbf{B} = \mathbf{A}^\Phi$. For $\zeta < \alpha$, let $u_\zeta \in B$ defined by $u_\zeta(\varphi) = \varphi(\zeta)$. Let $\mathbf{F} = \mathbf{B} \restriction [U]$, where $U = \{u_\zeta \mid \zeta < \alpha\}$. Use the result of Problem **4** to prove that **F** is a free $\hat{\mathbf{A}}$-algebra on the generating set U.

(b) Let $\mathfrak{B}$ be the class of all Boolean algebras. Show that $\mathfrak{B} = \hat{\mathbf{2}}$, and deduce from (a) that $\mathbf{2}^{2^n}$ is a free Boolean algebra on n generators.

6. (a) Let $\mathfrak{A}$ be a variety of algebras of type τ. Suppose that α is any cardinal number. Prove that an algebra **A** of type τ is a member of $\mathfrak{B}(\mathfrak{E}_\alpha(\mathfrak{A}))$ if and only if every subalgebra of **A** which is generated by no more than α elements is in $\mathfrak{A}$. (*Hint:* the proof that if $\mathbf{A} \in \mathfrak{B}(\mathfrak{E}_\alpha(\mathfrak{A}))$, then $\mathfrak{A}$ contains every subalgebra of **A** generated by $\leqq \alpha$ elements can be based on the result of Problem **4**.)

(b) Let $\mathfrak{A}$ be the variety of all Abelian groups. Let $\tau = \{\langle 0, 2\rangle, \langle 1, 0\rangle, \langle 2, 1\rangle\}$ be the type of Abelian groups. Suppose that $\mathbf{A} \in \mathfrak{A}_2$ (that is, $2x = 0$ for all $x \in A$). Define $B = (A - \{0\})^2 \cup \{0\}$, where 0 is the zero element of **A**.

Introduce a binary operation $\oplus$ on B by the rules

$$
\begin{array}{llll}
p \oplus 0 = 0 \oplus p = p & \text{for} & p \in B, & \\
p \oplus p = 0 & \text{for} & p \in B, & \\
\langle x, y\rangle \oplus \langle x, z\rangle = \langle x, y + z\rangle & \text{for} & y \neq z \quad \text{in} \quad A - \{0\}, & \\
\langle x, z\rangle \oplus \langle y, z\rangle = \langle x + y, z\rangle & \text{for} & x \neq y \quad \text{in} \quad A - \{0\}, & \\
\langle x, y\rangle \oplus \langle z, w\rangle = \langle x + z, y + w\rangle & \text{for} & x \neq z \quad \text{in} \quad y \neq w \quad \text{in} \quad A - \{0\}. &
\end{array}
$$

Let $\mathbf{B}$ be the algebra of type τ with this binary operation, the zero-ary operation that distinguishes 0, and the unary operation I_B. Show that $\mathbf{B}$ fails to satisfy the associative law provided A has at least four elements. However, prove that every subalgebra of $\mathbf{B}$ that is generated by two (or fewer) elements is an Abelian group. Use this example and **(a)** to conclude that $r(\mathfrak{A}) = 3$.

(c) Prove that $r(\mathfrak{A}) = 3$ if $\mathfrak{A}$ is the variety of all associative rings (*Hint:* define a multiplication on the example of **(b)** by the law $x \cdot y = 0$ for all x and y.)

7. Let $\mathfrak{A}$ be a variety of algebras of finitary type τ. Prove that an algebra $\mathbf{A}$ of type τ belongs to $\mathfrak{A}$ if and only if every finitely generated subalgebra of $\mathbf{A}$ is in $\mathfrak{A}$.

8. It follows from 3.2 (a) that if $\mathfrak{A}$ is a class of algebras of finitary type, then $r(\mathfrak{A}) \leqq \omega$. The purpose of this problem is to show that $r(\mathfrak{A})$ may be infinite even though $\sup \{|\tau(\xi)| \mid \xi < \rho\}$ is finite.

Let $\tau = \{\langle 0, 2\rangle, \langle 1, 0\rangle\}$. Define $A = \{0, e, b_0, b_1, c, d_0, d_1\}$, where the distinguished element $F_1(0)$ is 0. Define the binary operation F_0 on A by the conditions

$$
\begin{aligned}
F_0(\langle c, e\rangle) &= c, \\
F_0(\langle c, b_j\rangle) &= d_j, j < 2, \\
F_0(\langle d_j, e\rangle) &= d_j, j < 2, \\
F_0(\langle d_j, b_k\rangle) &= d_j, j < 2, k < 2, \\
F_0(\langle x, y\rangle) &= 0 \text{ in all other cases.}
\end{aligned}
$$

It is convenient to use the notation xy for $F_0(\langle x, y\rangle)$, and more generally $x_{i_0}x_{i_1} \cdots x_{i_{k-1}}$ for $(\cdots((x_{i_0}x_{i_1})x_{i_2}) \cdots)x_{i_{k-1}}$, and similar notation in the word algebra of type τ.

(a) Prove that $\mathbf{A} = \langle A; F_0, F_1\rangle$ satisfies the following identities

$$
\begin{aligned}
&A\colon \ \mathbf{0x}_0 = \mathbf{0},\ \mathbf{x}_0\mathbf{0} = \mathbf{0},\ \mathbf{x}_0(\mathbf{x}_1\mathbf{x}_2) = \mathbf{0}, \\
&B_n\colon \ \mathbf{x}_0\mathbf{x}_1 \cdots \mathbf{x}_{n-1}\mathbf{x}_0 = \mathbf{0},\ 1 \leqq n < \omega, \text{ and} \\
&C_n\colon \ \mathbf{x}_0\mathbf{x}_1 \cdots \mathbf{x}_{n-1}\mathbf{x}_1 = \mathbf{x}_0\mathbf{x}_1 \cdots \mathbf{x}_{n-1},\ 2 \leqq n < \omega.
\end{aligned}
$$

Call a word $\mathbf{w}$ in $W_{\tau n}$ *standard* if

$$\mathbf{w} = \mathbf{x}_{i_0}\mathbf{x}_{i_1} \cdots \mathbf{x}_{i_j},$$

where $j < n$ and $i_0, i_1, \cdots, i_j$ are distinct.

(b) Show that if $\mathbf{D}$ is an algebra of type τ that satisfies A, B_m and C_m for all $m < n$, then for any $\mathbf{w} \in W_{\tau n}$ either the identity $\mathbf{w} = \mathbf{0}$ is satisfied in $\mathbf{D}$, or there is a standard word $\mathbf{v} \in W_{\tau n}$ such that $\mathbf{w} = \mathbf{v}$ is satisfied in $\mathbf{D}$.

(c) Prove that if $\mathbf{v}_0$ and $\mathbf{v}_1$ are standard words in $W_{\tau n}$ such that $\mathbf{v}_0 = \mathbf{v}_1$ is satisfied in $\mathbf{A}$, then $\mathbf{v}_0$ and $\mathbf{v}_1$ are identically the same. Show that if $\mathbf{v}$ is standard in $\mathbf{W}_{\tau n}$, then $\mathbf{v} = \mathbf{0}$ is not satisfied in $\mathbf{A}$.

(d) Let $D = \{0, a, b_0, b_1, \cdots, b_{n-1}, c_0, c_1, \cdots, c_{n-1}\}$, and define a binary operation $F_0(\langle x, y\rangle) = xy$ by the conditions

$$\begin{aligned} ab_0 &= c_0 \\ c_i b_j &= c_j \quad \text{for } j \leqq i < n, \\ c_i b_{i+1} &= c_{i+1} \quad \text{for } i < n-1, \\ c_{n-1} a &= c_{n-1}, \quad \text{and} \\ xy &= 0 \text{ for all other choices of } x \text{ and } y \text{ in } D, \end{aligned}$$

and define the zero-ary operation F_1 by $F_1(0) = 0$. Prove that $\mathbf{D} = \langle D; F_0, F_1\rangle$ satisfies A, B_m, and C_m for all $m < n - 1$, but $\mathbf{D}$ does not satisfy B_{n-1}.

(e) Use (**d**) to prove that $\hat{\mathbf{A}}$ is properly contained in $\mathfrak{B}(\mathfrak{E}_{n-1}(\{\mathbf{A}\}))$ for all $n > 0$. Hence, $r(\hat{\mathbf{A}}) = \omega$.

9. Let α be an infinite cardinal number. An α-field of sets is a collection $\mathscr{F}$ of subsets of a set X with the properties that $Y \in \mathscr{F}$ implies $X - Y \in \mathscr{F}$, and $\{Y_i \mid i \in I\} \subseteq \mathscr{F}$, $|I| \leqq \alpha$ implies $\bigcup_{i \in I} Y_i \in \mathscr{F}$. Note that every α-field of sets can be considered as an α-complete Boolean algebra (see Example 4.1.8). An α-complete Boolean algebra $\mathbf{B}$ is called *α-representable* if $\mathbf{B}$ is a homomorphic image (preserving the α-ary joins and meets as well as the complement) of an α-field of sets. By Example 3.9 every countably complete Boolean algebra is $\aleph_0$-representable. The point of this problem is that $\aleph_0$ cannot be replaced in this statement by an arbitrary cardinal number α.

(a) Prove that $\hat{\mathbf{2}}$ is the class of all α-representable Boolean algebras (where $\mathbf{2}$ is considered as an α-complete Boolean algebra).

(b) Show that if $\alpha \geqq 2^{\aleph_0}$, and $\mathbf{B}$ is an α-representable Boolean algebra, then $\mathbf{B}$ satisfies the following condition: if $\{a_n \mid n < \omega\}$ is any countable subset of B, and the element b_φ is defined for each $\varphi \in 2^\omega$ to be the greatest lower bound of the set $\{a_n \mid \varphi(n) = 0\} \cup \{a_n' \mid \varphi(n) = 1\}$, then the least upper bound of $\{b_\varphi \mid \varphi \in 2^\omega\}$ is the unit element of $\mathbf{B}$.

(c) Let X be a Hausdorff topological space. For any set $A \subseteq X$, denote by A^- the closure of A in X, and let $A^c = X - A$. Finally, designate the interior of the set A by A^0; thus, $A^0 = A^{c-c}$. An open set A in X is called regular if $A^{-0} = A$. Prove that the collection of all regular open sets in X is a complete Boolean algebra with the operations $A \vee B = (A \cup B)^{-0}$, $A \wedge B = A \cap B$, $A' = A^{-0}$, and the least upper bound of $\{A_i \mid i \in I\}$ is given by $(\bigcup_{i \in I} A_i)^{-0}$.

(d) Prove that if X is any separable metric space without isolated points, then the Boolean algebra of regular open sets of X is $2^{\aleph_0}$-complete, but not $2^{\aleph_0}$-representable.

10. It is well known that every subgroup of a free group is free. This result is not a general property of free algebras. Using the result of Problem **5(b)**, it is easy to see that free Boolean algebras with $n \geqq 2$ generators have subalgebras that are not free. The purpose of this problem is to show that any subalgebra $\mathbf{A}$ of a word algebra $\mathbf{W}_{\tau\alpha}$ is completely free on a suitable set of generators. Define M to be the set of all $\mathbf{w} \in A$ such that either $\lambda(\mathbf{w}) = 1$, or else in the canonical representation

$$\mathbf{w} = \mathbf{\Phi}_{\xi}(\mathbf{v}_0\mathbf{v}_1 \cdots), \qquad \mathbf{v}_i \in W_{\tau\alpha}, \qquad \lambda(\mathbf{v}_i) < \lambda(\mathbf{w}),$$

at least one of the $\mathbf{v}_i$ is not in A.

(a) Prove that $[M] = A$. (*Hint:* show that the existence of an element of minimal length in $A - [M]$ leads to a contradiction.)

(b) Let $|M| = \beta$, and let φ be a homomorphism of $\mathbf{W}_{\tau\beta}$ to $\mathbf{A}$ which maps $\{\mathbf{x}_\zeta \mid \zeta < \beta\}$ one-to-one onto M. Prove that φ is an isomorphism. (*Hint:* if φ is not one-to-one, then among the pairs $\langle \mathbf{v}, \mathbf{w} \rangle$ such that $\varphi(\mathbf{v}) = \varphi(\mathbf{w})$ there is one for which $\lambda(\mathbf{v})$ is minimal. Use 1.3 and 1.5 to prove that this leads to a contradiction.)

(c) Deduce that $\mathbf{A}$ is completely free on the generating set M.

By using formal languages that contain expressions other than identities, it is possible to characterize classes of algebras that are more general than varieties. The purpose of the remaining problems is to sketch the proof of one result of this type.

In the following problems, $\tau \in \mathrm{Ord}^\rho$ is a fixed type, and α designates an infinite, regular cardinal number, that satisfies $\alpha > |\tau(\xi)|$ for all $\xi < \rho$. The *formal propositional language of type τ in α variables* is the class $L_{\tau\alpha}$ of sequences of symbols belonging to an alphabet $T_{\tau\alpha} = S_{\tau\alpha} \cup \{=, \wedge, \vee, \sim\}$ (where $S_{\tau\alpha}$ denotes the alphabet of the word algebra $W_{\tau\alpha}$), which is minimal subject to the conditions (i), (ii), and (iii) listed below. If $\mathbf{s}$ is a sequence of elements in $T_{\tau\alpha}$ indexed by W, we denote the set $\{\zeta < \alpha \mid x_\zeta \in \mathbf{s}(W)\}$ by Var $\mathbf{s}$. It is clear that if $\mathbf{s}$ is equivalent to $\mathbf{t}$, then Var $\mathbf{s}$ = Var $\mathbf{t}$. Therefore we can consider "Var" as being defined on sequences.

(i) If $\mathbf{v}$ and $\mathbf{w}$ are in $W_{\tau\gamma}$ for some $\gamma < \alpha$, then $(\mathbf{v} = \mathbf{w})$ is a member of $L_{\tau\alpha}$.

(ii) If $\{\mathbf{f}_\nu \mid \nu < \mu\} \subseteq L_{\tau\alpha}$ (μ an ordinal), and $\bigcup_{\nu<\mu} \mathrm{Var}\, \mathbf{f}_\nu \subseteq \gamma$ for some $\gamma < \alpha$, then

$$(\mathbf{f}_0 \wedge \mathbf{f}_1 \wedge \cdots \wedge \mathbf{f}_\nu \wedge \cdots) \in L_{\tau\alpha}, \qquad \text{and} \qquad (\mathbf{f}_0 \vee \mathbf{f}_1 \vee \cdots \vee \mathbf{f}_\nu \vee \cdots) \in L_{\tau\alpha}.$$

(iii) If $\mathbf{f} \in L_{\tau\alpha}$, then $\sim(\mathbf{f}) \in L_{\tau\alpha}$.

(*Remark.* The symbols $\wedge$, $\vee$, and $\sim$ are to be interpreted as the usual propositional connectives "and," "or," and "not," except that $\wedge$ and $\vee$ are allowed to act as infinitary operators.)

11. Prove that

(a) there is a smallest class $L_{\tau\alpha}$ with the properties (i), (ii), and (iii).

(b) If $\mathbf{f} \in L_{\tau\alpha}$, then there exists $\gamma < \alpha$ such that $\operatorname{Var} \mathbf{t} \subseteq \gamma$.

(c) If $\mathbf{f} \in L_{\tau\alpha}$, then either

(i) $\mathbf{f} = (\mathbf{v} = \mathbf{w})$, where $\mathbf{v}$ and $\mathbf{w}$ are in $W_{\tau\gamma}$ for some $\gamma < \alpha$;

(ii) $\mathbf{f} = (\mathbf{f}_0 \wedge \mathbf{f}_1 \wedge \cdots \wedge \mathbf{f}_\nu \wedge \cdots)$, $\nu < \mu$, where all $\mathbf{f}_\nu$ are in $L_{\tau\alpha}$ and $\lambda(\mathbf{f}_\nu) < \lambda(\mathbf{f})$;

(iii) $\mathbf{f} = (\mathbf{f}_0 \vee \mathbf{f}_1 \vee \cdots \vee \mathbf{f}_\nu \vee \cdots)$, $\nu < \mu$, where all $\mathbf{f}_\nu$ are in $L_{\tau\alpha}$ and $\lambda(\mathbf{f}_\nu) < \lambda(\mathbf{f})$; or

(iv) $\mathbf{f} = \sim(\mathbf{g})$, where $\mathbf{g} \in L_{\tau\alpha}$ and $\lambda(\mathbf{g}) < \lambda(\mathbf{f})$.

(d) The representations given in **(c)** are unique.

The notion of an algebra satisfying an identity must be generalized to the formulas of $L_{\tau\alpha}$. In order to do this, we need some preliminary results. If **A** is an algebra of type τ, then each element $\mathbf{a} \in A^\alpha$ determines a unique homomorphism $\varphi_\mathbf{a}$ of $\mathbf{W}_{\tau\alpha}$ to **A** satisfying $\varphi_\mathbf{a}(\mathbf{x}_\zeta) = \mathbf{a}(\zeta)$ for all $\zeta < \alpha$ (by Theorem 1.6).

12. Prove that for each algebra **A** of type τ, and each $\mathbf{f} \in L_{\tau\alpha}$, there is precisely one α-ary relation $R(\mathbf{A}, \mathbf{f})$ on **A** satisfying the conditions:

(i) If $\mathbf{f} = (\mathbf{v} = \mathbf{w})$, then $R(\mathbf{A}, \mathbf{f}) = \{\mathbf{a} \in A^\alpha \mid \varphi_\mathbf{a}(\mathbf{v}) = \varphi_\mathbf{a}(\mathbf{w})\}$.

(ii) If $\mathbf{f} = (\mathbf{f}_0 \wedge \mathbf{f}_1 \wedge \cdots \wedge \mathbf{f}_\nu \wedge \cdots)$, $\nu < \mu$, where all $\mathbf{f}_\nu$ are in $L_{\tau\alpha}$ and $\lambda(\mathbf{f}_\nu) < \lambda(\mathbf{f})$, then $R(\mathbf{A}, \mathbf{f}) = \bigcap_{\nu<\mu} R(\mathbf{A}, \mathbf{f}_\nu)$.

(iii) If $\mathbf{f} = (\mathbf{f}_0 \vee \mathbf{f}_1 \vee \cdots \vee \mathbf{f}_\nu \vee \cdots)$, $\nu < \mu$, where all $\mathbf{f}_\nu$ are in $L_{\tau\alpha}$ and $\lambda(\mathbf{f}_\nu) < \lambda(\mathbf{f})$, then $R(\mathbf{A}, \mathbf{f}) = \bigcup_{\nu<\mu} R(\mathbf{A}, \mathbf{f}_\nu)$.

(iv) If $\mathbf{f} = \sim(\mathbf{g})$, where $\mathbf{g} \in L_{\tau\alpha}$ and $\lambda(\mathbf{g}) < \lambda(\mathbf{f})$, then $R(\mathbf{A}, \mathbf{f}) = A^\alpha - R(\mathbf{A}, \mathbf{g})$.

13. Prove that if **A** is an algebra of type τ, $\mathbf{f} \in L_{\tau\alpha}$, and **a** and **b** are elements of A^α such that $\mathbf{a} \restriction \operatorname{Var} \mathbf{f} = \mathbf{b} \restriction \operatorname{Var} \mathbf{f}$, then $\mathbf{a} \in R(\mathbf{A}, \mathbf{f})$ if and only if $\mathbf{b} \in R(\mathbf{A}, \mathbf{f})$.

14. Let **A** and **B** be algebras of type τ, and suppose that φ is an isomorphism of **A** to **B**. Prove that for each $\mathbf{f} \in L_{\tau\alpha}$, $\varphi \circ R(\mathbf{A}, \mathbf{f}) = R(\mathbf{B}, \mathbf{f})$.

15. Let **A** be an algebra of type τ, and suppose that **B** is a subalgebra of **A**. Prove that for each $\mathbf{f} \in L_{\tau\alpha}$, $R(B, \mathbf{f}) = R(\mathbf{A}, \mathbf{f}) \cap B^\alpha$.

If $\mathbf{f} \in L_{\tau\alpha}$ and **A** is an algebra of type τ, we say that **A** *satisfies* **f** (or **f** *is satisfied in* **A**) if $R(\mathbf{A}, \mathbf{f}) = A^\alpha$. It is easy to see that if $\mathbf{f} = (\mathbf{v} = \mathbf{w})$, then this definition is equivalent to the concept introduced in Definition 2.1.

16. Let **A** be an algebra of type τ, and suppose that $\mathbf{f} \in L_{\tau\alpha}$ is such that if $X \subseteq A$ and $|X| < \alpha$, then $\mathbf{A} \restriction [X]$ satisfies **f**. Use the result of Problem **11(b)** to prove that **A** satisfies **f**.

Let γ be a fixed cardinal number $<\alpha$. Let **A** be an algebra of type τ. Corresponding to γ and **A**, define $\mathbf{g}_{\gamma\mathbf{A}} \in L_{\tau\alpha}$ as follows: well-order the set $W^2_{\tau\gamma}$, say $W^2_{\tau\gamma} = \{\langle \mathbf{v}_\nu, \mathbf{w}_\nu \rangle \mid \nu < \mu\}$. Also, well-order A^α, say $A^\alpha = \{\mathbf{a}_\kappa \mid \kappa < \lambda\}$.

For $\nu < \mu$ and $\kappa < \lambda$, define $\mathbf{k}_{\kappa\nu} = (\mathbf{v}_\nu = \mathbf{w}_\nu)$ if $\varphi_{\mathbf{a}_\kappa}(\mathbf{v}_\nu) = \varphi_{\mathbf{a}_\kappa}(\mathbf{w}_\nu)$, and $\mathbf{k}_{\kappa\nu} = \sim(\mathbf{v}_\nu = \mathbf{w}_\nu)$ if $\varphi_{\mathbf{a}_\kappa}(\mathbf{v}_\nu) \neq \varphi_{\mathbf{a}_\kappa}(\mathbf{w}_\nu)$. Let $\mathbf{h}_\kappa = (\mathbf{k}_{\kappa 0} \wedge \mathbf{k}_{\kappa 1} \wedge \cdots \wedge \mathbf{k}_{\kappa\nu} \wedge \cdots)$, $(\nu < \mu)$, and $\mathbf{g}_{\gamma\mathbf{A}} = (\mathbf{h}_0 \vee \mathbf{h}_1 \vee \cdots \vee \mathbf{h}_\kappa \vee \cdots)$, $(\kappa < \lambda)$.

17. Prove that

(a) $\mathbf{A}$ satisfies $\mathbf{g}_{\gamma\mathbf{A}}$.

(b) If $\mathbf{B}$ is an algebra of type τ and $\mathbf{b} \in B^\gamma$ is such that

(i) $[\{\mathbf{b}(\zeta) \mid \zeta < \gamma\}] = B$, and

(ii) there is a $\mathbf{c} \in B^\alpha$ with $\mathbf{c} \restriction \gamma = \mathbf{b}$ and $\mathbf{c} \in R(\mathbf{B}, \mathbf{g}_{\gamma\mathbf{A}})$,

then $\mathbf{B}$ is isomorphic to a subalgebra of $\mathbf{A}$.

Let $\mathfrak{A}$ be a class of algebras of type τ such that $\mathscr{S}\mathfrak{A} = \mathfrak{A}$. Let $\gamma < \alpha$. Select a well-ordered set of algebras $\{\mathbf{A}_\xi \mid \xi < \nu\}$ (from among the homomorphic images of $\mathbf{W}_{\tau\alpha}$, for example), such that every $\mathbf{A}_\xi$ is in $\mathfrak{A}$, and if $\mathbf{B} \in \mathfrak{A}$ is generated by a set of cardinality $\leqq \gamma$, then $\mathbf{B}$ is isomorphic to a subalgebra of some $\mathbf{A}_\xi$. For each $\xi < \nu$, let $\mathbf{g}_\xi$ be the formula $\mathbf{g}_{\gamma A_\xi}$ constructed in the previous problem. Define

$$\mathbf{f}_\gamma = \langle(\mathbf{g}_0 \vee \mathbf{g}_1 \vee \cdots \vee \mathbf{g}_\xi \vee \cdots)\rangle_{\xi<\gamma}.$$

18. Prove that every $\mathbf{A} \in \mathfrak{A}$ satisfies $\mathbf{f}_\gamma$, and that if $\mathbf{B}$ is an algebra of type τ which is generated by a set of cardinality $\leqq \gamma$, such that $\mathbf{B}$ satisfies $\mathbf{f}_\gamma$, then $\mathbf{B} \in \mathfrak{A}$.

19. Use the results of Problems **14–18** to prove the following:

Theorem. A necessary and sufficient condition for a class $\mathfrak{A}$ of algebras to be the class of all algebras which satisfy all of the formulas in a subset $\mathscr{F}$ of $L_{\tau\alpha}$ is that $\mathscr{S}\mathfrak{A} = \mathscr{S}_\alpha^{-1}\mathfrak{A} = \mathfrak{A}$.

(Recall that $\mathscr{S}_\alpha^{-1}\mathfrak{A}$ is the class of all algebras $\mathbf{A}$ of type τ such that $\mathbf{A} \restriction [X] \in \mathfrak{A}$ for all $X \subseteq A$ such that $|X| < \alpha$.)

20. Let β be an infinite cardinal number. A β-complete Boolean algebra $\mathbf{B}$ is called *β-distributive* if $\bigwedge_{\xi<\beta}(\bigvee_{\eta<\beta} a_{\xi\eta}) = \text{l.u.b.}\{\bigwedge_{\xi<\beta} a_{\xi\varphi(\xi)} \mid \varphi \in \beta^\beta\}$ for every set $\{a_{\xi\eta} \mid \xi < \beta, \eta < \beta\} \subseteq B$.

(a) Show that every β-field of sets is β-distributive. Conversely, prove that a β-distributive Boolean algebra that is generated by a set of cardinality $\leqq \beta$ is isomorphic to a β-field. (*Hint:* Let $\{x_\xi \mid \xi < \beta\}$ be a set of generators of $\mathbf{B}$. Define $a_{\xi 0} = x_\xi$, $a_{\xi 1} = x'_\xi$ for $\xi < \beta$, and $b_\varphi = \bigwedge_{\xi<\beta} a_{\xi\varphi(\xi)}$ for $\varphi \in 2^\beta$. Prove that the mapping $c \to \{\varphi \mid b_\varphi \leqq c\}$ is an isomorphism of $\mathbf{B}$ (considered as a β-complete Boolean algebra) onto a β-field of subsets of 2^β.)

(b) Let $\mathbf{B}$ be a β-complete Boolean algebra. Prove that $\mathbf{B}$ is β-distributive if and only if the following condition is satisfied for every set

$$\{a_{\xi\eta} \mid \xi < \beta, \eta < \beta\} \subseteq B:$$

(i) if $\bigwedge_{\xi<\beta} a_{\xi\varphi(\xi)} = 0$ for all $\varphi \in \beta^\beta$, then $\bigwedge_{\xi<\beta}(\bigvee_{\eta<\beta} a_{\xi\eta}) = 0$.

Show that this condition can be expressed by a formula of $L_{\tau\alpha}$, where τ is the type of β-complete Boolean algebras and α is the smallest cardinal number exceeding β. (*Hint:* To prove that (i) implies β-distributivity, use the distributive law mentioned in Problem **16** of Chapter 4.)

(c) Use the results of (**a**) and (**b**), together with the theorem of Problem **19**, to show that the following conditions are equivalent for a β-complete Boolean algebra **B**.

(i) **B** is β-distributive.

(ii) Every subalgebra of **B** generated by a set of cardinality $\leqq \beta$ is β-distributive.

(iii) Every subalgebra of **B** generated by a set of cardinality $\leqq \beta$ is isomorphic to a β-field of sets.

(d) Use the result of (**c**) to prove that every β-distributive Boolean algebra is β-representable. (*Hint:* Note that the class $\mathfrak{R}_\beta$ of all β-representable Boolean algebras is a variety by Problem **9**, and therefore $\mathscr{S}_\alpha^{-1}\mathfrak{R}_\beta = \mathfrak{R}_\beta$ by Proposition 3.7.)

(e) Show that the class of all β-distributive Boolean algebras is not a variety.

Notes to Chapter 5

Birkhoff's theorem first appeared in [1]. The generalization to infinitary algebras was made by Slominski [47]. Languages with infinitely long sentences have been studied systematically by Karp [31].

The axiom rank of varieties is discussed in Chapter 4 of Cohn's book. Most of the results in our Section 3 can be found there, as well as the results given in Problems **1**, **3**, **5**, **6**, and **7**. The axiom rank of the familiar varieties of algebras is finite, since the majority of algebraic systems are defined by finite sets of identities. However, it frequently requires ingenuity to find the exact value of $r(\mathfrak{A})$. The example given in Problem **8** is due to Lyndon [36].

The idea of deducing the Loomis-Sikorski theorem from Birkhoff's result is due to Tarski and Scott (see [53]). Proposition 3.7 is related to the compactness theorem of mathematical logic (see [9], p. 213). In fact, the finite version of 3.7 (that is, under the assumption that τ is finitary, **A** is in $\mathfrak{A}$ if and only if every finite subsystem of **A** belongs to $\mathfrak{A}$) is easily derivable from the compactness theorem.

The problem of Characterizing the subalgebras of a free algebra has attracted much attention. Probably the most celebrated result in this direction is Schreier's theorem that every subgroup of a free group is free. Problem **10**, which is due to Feigelstock [18], provides another instance in which a subalgebra of a free algebra is free. Nevertheless, it appears that this phenomenon is exceptional. There are many varieties $\mathfrak{A}$ for which little more is known about the subalgebras of the free $\mathfrak{A}$-algebras than the fact that they need not be free. Such is the case for example when $\mathfrak{A}$ is the class of all lattices or the class of all Boolean algebras.

The theorem of Problem **19** is a modification of a result due to Tarski [52]. This theorem is as close as we come in this book to the subject of model theory. For an introduction to this active field, Tarski's paper [51], Robinson's work [43], and Cohn's book [9] are recommended.

Bibliography

This is not intended to be a research bibliography. The only papers and books listed here are those to which reference has been made in the text, and these references have been limited to works in the English language. More complete bibliographies can be found in the books of Cohn [9] and Gratzer [21].

[1] Birkhoff, G. (1935), On the structure of abstract algebras, *Proc. Cambridge Phil. Soc.*, **31**, 433–454.

[2] ——— (1944), Subdirect unions in universal algebras, *Bull. Amer. Math. Soc.*, **50**, 764–768.

[3] ——— (1948), *Lattice Theory*, Amer. Math. Soc. Colloquium Publications, Vol. 25, Revised ed., New York.

[4] ———, and O. Frink (1948), Representations of lattices by sets, *Trans. Amer. Math. Soc.*, **64**, 299–316.

[5] ——— and R. S. Pierce (1956), Lattice ordered rings, *Anais da Acad. Bras. de Ciencias*, **28**, 41–69.

[6] Cartan, H. and S. Eilenberg (1956), *Homological Algebra*, Princeton University Press, Princeton, N.J.

[7] Chang, C. C., B. Jónsson, and A. Tarski (1964), Refinement properties for relational structures, *Fund. Math.*, **55**, 249–281.

[8] Christensen, D. J. and R. S. Pierce (1959), Free products of α-distributive Boolean algebras, *Math. Scand.* **7**, 81–105.

[9] Cohn, P. M. (1965), *Universal Algebra*, Harper & Row, New York.

[10] Crawley, P. (1960), Lattices whose congruences form a Boolean algebra, *Pac. Jour. of Math.*, **10**, 787–795.

[11] ——— and B. Jónsson (1964), Refinements for infinite direct decompositions of algebraic systems, *Pac. Jour. of Math.*, **14**, 797–855.

[12] Dean, R. A. (1956), Completely free lattices generated by partially ordered sets, *Trans. Amer. Math. Soc.*, **83**, 238–249.

[13] Dilworth, R. P. (1945), Lattices with unique complements, *Trans. Amer. Math Soc.*, **57**, 123–154.

[14] ——— (1950), The structure of relatively complemented lattices, *Annals of Math.*, **51**, 348–359.

[15] ——— (1961), Structure and decomposition theory of lattices, *Proc. of Symp. in Pure Math.*, II, (Providence), 3–16.

[16] ——— and P. Crawley (1960), Decomposition theory for lattices without chain conditions, *Trans. Amer. Math. Soc.*, **97**, 1–22.

[17] Dwinger, P. and F. M. Yaqub (1963), Generalized free products of Boolean algebras with an amalgamated subalgebra, *Indag. Math.*, **25**, 225–231.

[18] Feigelstock, S. (1965), A universal subalgebra theorem, *Amer. Math. Monthly*, **72**, 884–888.

[19] Frayne, T., A. C. Morel, and D. S. Scott (1962), Reduced direct products, *Fund. Math.*, **51**, 195–228.

[20] Gödel, K. (1940), *The Consistency of the Continuum Hypothesis*, Princeton University Press, Princeton, N.J.

[21] Grätzer, G. (1967), *Universal Algebra*, Van Nostrand, Princeton, N.J.

[22] ——— and E. T. Schmidt, (1960), On inaccessible and minimal congruence relations, I, *Acta Sci. Math. Szeged*, **21**, 337–342.

[23] ——— and ——— (1963), Characterizations of congruence lattices of abstract algebras, *Acta Sci. Math. Szeged*, **24**, 34–59.

[24] Halmos, P. R. (1961), Injective and projective Boolean algebras, *Proc. of Symp. in Pure Math.* **II**, (Providence), 114–122.

[25] Hashimoto, J. (1957), Direct, subdirect decompositions and congruence relations, *Osaka Math. Jour.*, **9**, 87–112.

[26] Holland, C. (1963), The lattice ordered group of automorphisms of an ordered set, *Mich. Jour. of Math.*, **10**, 399–408.

[27] Jónsson, B. (1956), Universal relational systems, *Math. Scand.*, **4**, 193–208.

[28] ——— (1957), On direct decompositions of torsion free abelian groups, *Math. Scand.*, **5**, 230–235.

[29] ——— and A. Tarski (1947), *Direct Decompositions of Finite Algebraic Systems*, University of Notre Dame Press, Notre Dame, Ind.

[30] ——— (1961), On two properties of free algebras, *Math. Scand.*, **9**, 95–101.

[31] Karp, C. R. (1964), *Languages with Expressions of Infinite Length*, North-Holland Publishing Company, Amsterdam.

[32] Kelley, J. L. (1955), *General Topology*, Van Nostrand, Princeton, N.J.

[33] Kochen, S. (1961), Ultraproducts in the theory of models, *Annals of Math.*, 74, 221–261.

[34] Kuroš, A. G. (1963), *Lectures on General Algebra*, Chelsea, New York.

[35] Loomis, L. (1947), On the representation of σ-complete Boolean algebras, *Bull. Amer. Math. Soc.*, **53**, 757–760.

[36] Lyndon, R. C. (1954), Identities in finite algebras, *Proc. Amer. Math. Soc.*, **5**, 8–9.

[37] Marczewski, E. (1958), A general scheme of the notions of independence in mathematics, *Bull. Acad. Polon. Sci.*, **6**, 731–736.

[38] ——— (1961), Independence and homomorphisms in abstract algebras, *Fund. Math.*, **50**, 45–61.

[39] Morley, M. and R. L. Vaught (1962), Homogeneous universal models, *Math. Scand.*, **11**, 37–57.

[40] Neumann, B. H. (1962), Universal Algebra, Lecture Notes, New York University.

[41] Ore, O. (1935 and 1936), On the foundations of abstract algebra, I, II, *Annals of Math.*, **36**, 406–437; **37**, 265–292.

[42] Pierce, R. S. (1963), A note on free products of abstract algebras, *Indag. Math.*, **25**, 401–407.

[43] Robinson, A. (1963), *Introduction to Model Theory and to the Metamathematics of Algebra*, North-Holland Publishing Company, Amsterdam.

[44] Sikorski, R. (1948), On the representation of Boolean algebras as fields of sets, *Fund. Math.*, **35,** 247–256.

[45] ——— (1948), A theorem on extension of homomorphisms, *Ann. Soc. Pol. Math.*, **21,** 332–335.

[46] ——— (1953), Products of abstract algebras, *Fund. Math.*, **39,** 211–228.

[47] Slominski, J. (1959), The theory of abstract algebras with infinitary operations, *Rozprawy Mat.*, **18,** 1–69.

[48] Stone, M. H. (1936), The theory of representations for Boolean algebras, *Trans. Amer. Math. Soc.*, **40,** 37–111.

[49] Swierczkowski, S. (1960), Algebras which are independently generated by every n elements, *Fund. Math.*, **49,** 93–104.

[50] ——— (1961), On isomorphic free algebras, *Fund. Math.*, **50,** 35–44.

[51] Tarski, A. (1950), Some notions and methods on the borderline of algebra and metamathematics, *Proc. International Congress of Mathematicians*, (Cambridge, Mass), 705–720.

[52] ——— (1954 and 1955), Contributions to the theory of models, I, II, III, *Indag. Math.*, **16,** 572–88; **17,** 56–64.

[53] ——— (1955), Metamathematical proofs for some representation theorems for Boolean algebras, *Bull. Amer. Math. Soc.*, **61** 523.

[54] Urbanik, K. (1959), A representation theorem for Marczewski's algebras, *Fund. Math.*, **48,** 147–167.

[55] ——— (1963), Remark on independence in finite algebras, *Coll. Math.*, **11,** 1–12.

[56] Yaqub, F. M. (1963), Free extensions of Boolean algebras, *Pac. Jour. of Math.*, **13,** 761–771.

Symbols

Index